HOLT McDOUGAL

Biology

Introducing Biology Unit 1 Resource Book

HOLT McDOUGAL
a division of Houghton Mifflin Harcourt

Printed in the United States of America.

ISBN-13: 978-0-618-72522-9 ISBN-10: 0-618-72522-9

4 5 6 7 8 9 - MJT - 12 11 10 09 08

Introducing Biology
Unit 1 Resource Book

CHAPTER 1 BIOLOGY IN THE 21ST CENTURY

CHAPTER 2 CHEMISTRY OF LIFE

Name Period Date

SECTION **1.1** | THE STUDY OF LIFE

Study Guide

KEY CONCEPT

Biologists study life in all its forms.

VOCABULARY		
biosphere	biology	metabolism
biodiversity	organism	DNA
species	cell	

MAIN IDEA: Earth is home to an incredible diversity of life.

Take notes about the diversity of life on Earth in the chart below.

1. In the box labeled **The Biosphere,** list examples of environments on earth in which life is found.
2. In the box labeled **Biodiversity,** write a definition of the term in your own words.
3. In the box labeled **Species,** briefly explain what a species is.

Earth is home to an incredible diversity of life.

→ **The Biosphere**

→ **Biodiversity**

→ **Species**

4. How is biodiversity related to the biosphere?

____

____

5. In general, how does biodiversity vary across the biosphere?

____

____

MAIN IDEA: All organisms share certain characteristics.

6. Before reading, take a quick look at the headings in this main idea. What are the four characteristics that identify something as living?

____

7. As you read, take notes on how the four basic characteristics help define what is a living thing.

Characteristic	Summary Sentence
Cells	
Energy and metabolism	
Response to environment	
Reproduction and development	

Vocabulary Check

8. The word *biosphere* is made up of two word parts: *bio-* and *sphere.* How can these two word parts help you to remember the definition of biosphere?

____

____

9. What is an organism?

____

10. The term *metabolism* is based on a Greek word that means "change." How is this meaning related to the meaning of metabolism?

____

11. How is DNA related to reproduction?

____

Name Period Date

SECTION 1.1 | THE STUDY OF LIFE

Power Notes

Biosphere:

Biodiversity:

Species:

Organism:

Characteristics Shared by Organisms	Details
1.	
2.	
3.	
4.	

SECTION 1.1 | THE STUDY OF LIFE

Reinforcement

KEY CONCEPT Biologists study life in all its forms.

Biology is the scientific study of all forms of life. Living things are found almost everywhere on Earth, from very hot environments to very cold environments and from the dry deserts to the ocean floor. The types of living things found in a particular region depend on which can survive there. Those living things that can survive in an environment can differ greatly in size and shape.

All of the living things on Earth, and all of the places in which they live, make up the **biosphere.** The variety of living things in a certain area, or across the entire biosphere, is called **biodiversity.** Biodiversity can be measured in terms of the number of species in an area or across the biosphere. Although there are several definitions of the term *species,* you can think of a **species** as a certain type of living things that can reproduce by interbreeding.

At least two million species exist on Earth. Each individual living thing, no matter the species, is an **organism.** Every organism, from any species, shares certain characteristics of life.

- All organisms are made of one or more cells. A **cell** is the basic unit of life on Earth.
- All organisms need chemical energy to carry out all of their cell functions. Energy is used for **metabolism,** which is all of the chemical processes that build up or break down materials.
- All organisms respond to physical factors, or stimuli, in their environment.
- Members of a species must be able to reproduce so that the species will survive. When organisms reproduce they pass on their genetic material, which is called **DNA,** to their offspring.

1. Where do all living things on Earth exist?

____

2. What is one way in which biodiversity can be measured?

____

____

3. List the four characteristics of living things.

____

____

4. What is metabolism?

____

____

Name Period Date

SECTION **1.2** | UNIFYING THEMES OF BIOLOGY

Study Guide

KEY CONCEPT

Unifying themes connect concepts from many fields of biology.

VOCABULARY		
system	homeostasis	adaptation
ecosystem	evolution	

MAIN IDEA: All levels of life have systems of related parts.

1. What is a system?

____

____

2. What are some examples of systems?

____

____

Complete the table by writing either the level of life or an example of a system at that level of life.

Level	Example
3.	Chemicals and processes interact in a precise way so that a cell can function properly.
Cells	**4.**
5.	Different parts of a living thing work together so that the living thing can survive.
Ecosystem	**6.**

MAIN IDEA: Structure and function are related in biology.

7. What are structure and function?

____

____

8. Give an example of how structure and function are related on the cellular level.

____

____

MAIN IDEA: Organisms must maintain homeostasis to survive in diverse environments.

9. What is homeostasis?

____

10. Why is homeostasis important to the survival of an organism?

____

____

11. In the space below, draw a sketch to help you remember what negative feedback is.

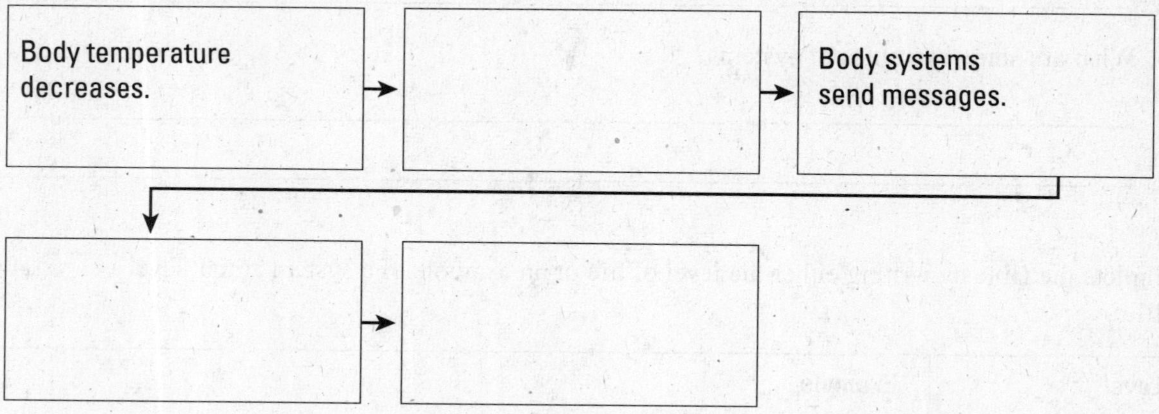

MAIN IDEA: Evolution explains the unity and diversity of life.

12. What is evolution?

____

13. Over the course of time, evolution ______________ the genetic makeup of a population.

14. ______________ are genetic traits that give an advantage to an individual and can be passed on to offspring.

Vocabulary Check

15. A system in which living and nonliving things in a certain area interact is called a(n) ______________.

16. The maintenance of constant internal conditions in an organism is called ______________.

SECTION 1.2 | UNIFYING THEMES OF BIOLOGY
Power Notes

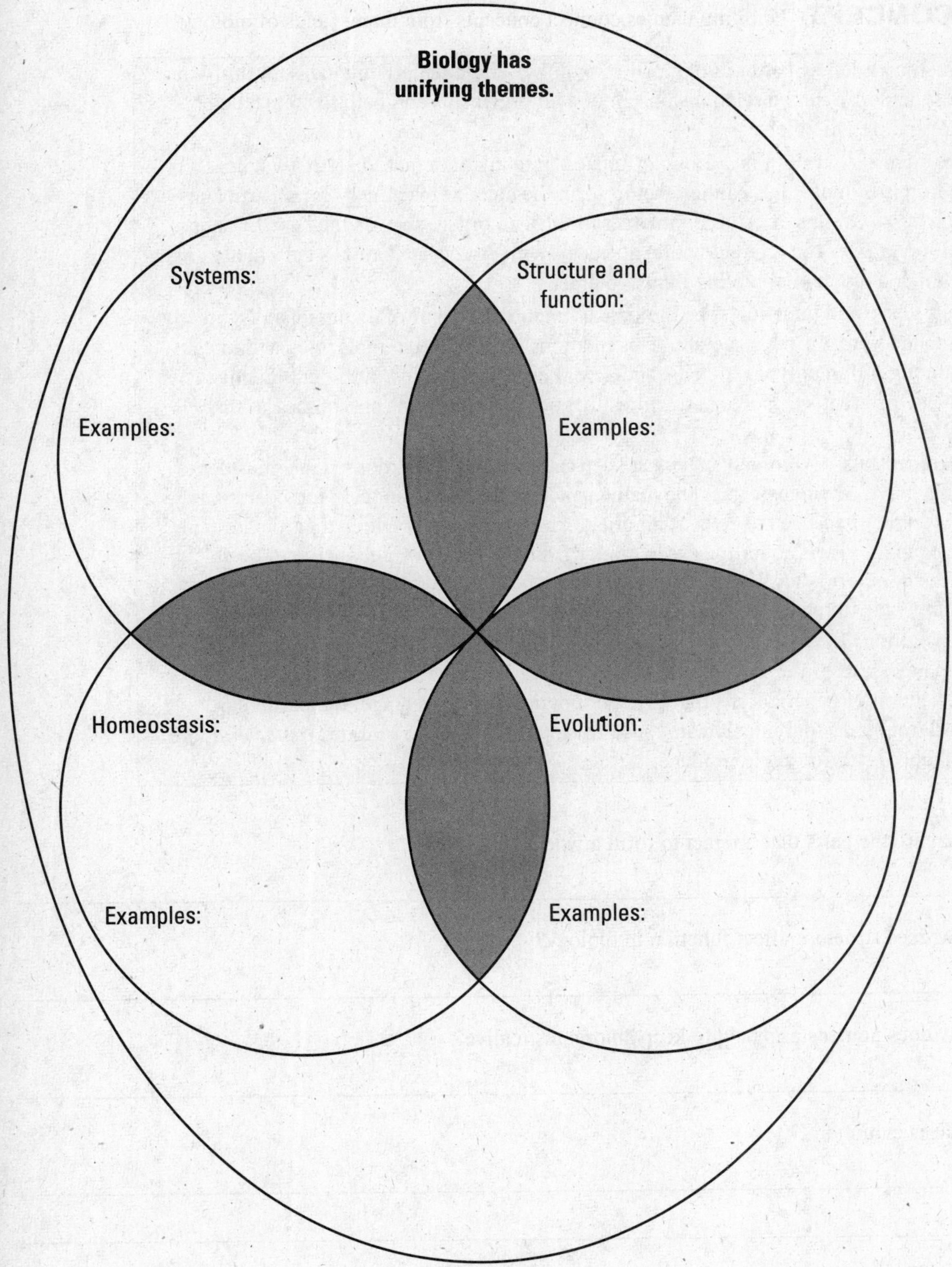

Name Period Date

SECTION 1.2 | UNIFYING THEMES OF BIOLOGY

Reinforcement

KEY CONCEPT Unifying themes connect concepts from many fields of biology.

Several major concepts run through all of biology. These underlying ideas, or unifying themes, demonstrate the relationships among all organisms and help to connect one field of biology to others.

- Systems: A **system** is a group of related parts that interact to form a whole. Groups of molecules can interact to form the cellular machinery for a particular process. Groups of cells can interact to form an organ, such as the heart. Groups of organisms can interact within an ecosystem. An **ecosystem** is a certain area in which living and nonliving things interact.
- Structure and Function: The biological function of a part of an organism is directly related to that part's structure. This relationship is found in molecules within cells, among different types of cells, and across different species. Different organisms have different specialized structures that perform functions specialized to that species.
- Homeostasis: All organisms must keep their internal conditions stable in order to stay alive. **Homeostasis** is the maintenance of these conditions. Homeostasis is necessary because the cells of all organisms function best within a particular range of conditions. If conditions vary too far from the ideal set, cells will not be able to function normally. When internal conditions change, a negative feedback system often acts to return the condition to normal.
- Evolution: **Evolution** is the process by which the genetic makeup of a population changes over time. One way in which evolution occurs is through the natural selection of genetic traits that give an individual an advantage in a particular environment. Individuals with these advantageous traits, or **adaptations,** are more likely to survive and reproduce.

1. What are the parts that interact to form a whole ecosystem?

____

2. How does structure affect function in biology?

____

3. How does homeostasis help to keep an organism alive?

____

4. What is evolution?

____

____

Name Period Date

SECTION
1.3

SCIENTIFIC THINKING AND PROCESSES

Study Guide

KEY CONCEPT

Science is a way of thinking, questioning, and gathering evidence.

VOCABULARY			
observation	hypothesis	independent variable	constant
data	experiment	dependent variable	theory

MAIN IDEA: Like all science, biology is a process of inquiry.

Complete the table below by giving a brief description and a brief example of each of the scientific process terms.

Scientific Process	Description	Example
Observation	1.	2.
Data	3.	4.
Hypothesis	5.	6.

7. How do scientists use statistics when they test a hypothesis?

____

____

8. Why is it important that a scientist's results are evaluated by other scientists?

____

____

Name Period Date

9. Look at Figure 1.10. Beginning with observation, what are the five parts of scientific thinking?

____

____

MAIN IDEA: Biologists use experiments to test hypotheses.

10. In ______________ studies, scientists do not interfere with what is going on.

11. Scientists can test hypotheses through ______________ .

12. A(n) ______________ variable is one which is observed and not manipulated by the scientist.

13. How are constants different from independent variables?

____

____

MAIN IDEA: A theory explains a wide range of observations.

14. What is the difference between a theory and a hypothesis?

____

15. Why are theories never proven?

____

Vocabulary Check

16. What is a hypothesis?

____

17. How can you remember the difference between an independent variable and a dependent variable? Think about what the words *independent* and *dependent* mean.

____

____

____

Name Period Date

SECTION **1.3** | SCIENTIFIC THINKING AND PROCESSES

Power Notes

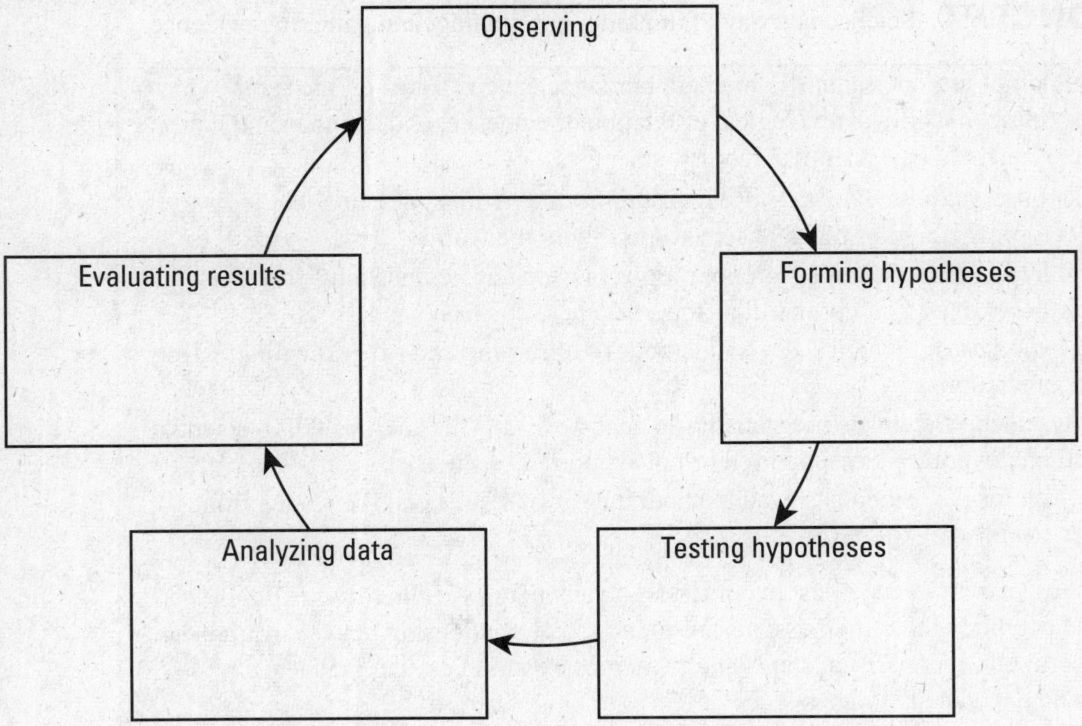

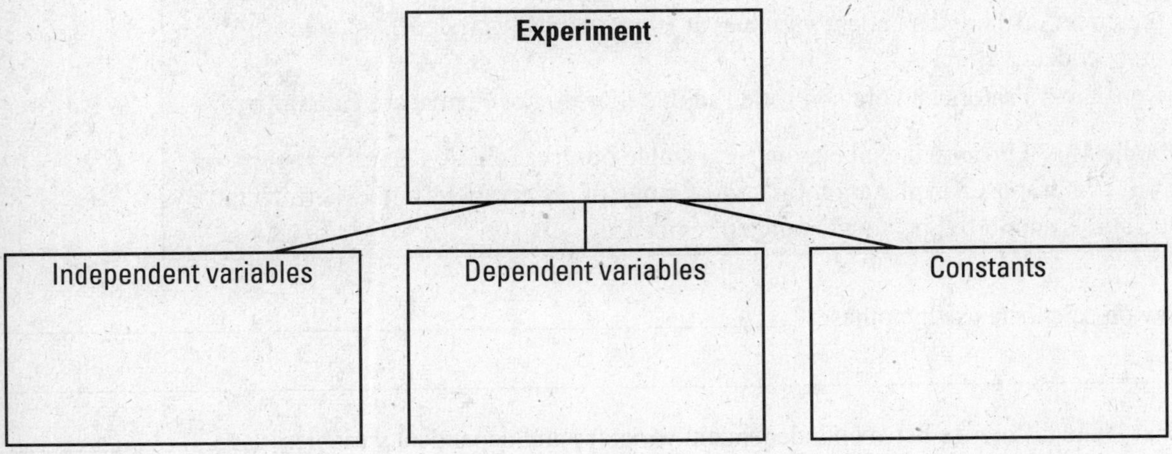

Theory:

Name Period Date

SECTION **1.3** | SCIENTIFIC THINKING AND PROCESSES

Reinforcement

KEY CONCEPT Science is a way of thinking, questioning, and gathering evidence.

Scientists do not use one scientific method, but all science is based on the same principles: curiosity, critical and logical evaluation of evidence, and the open and honest exchange of data. In any scientific inquiry, scientists

- Make observations: Scientists use their senses and various measurement tools to collect information, or make **observations,** about the world.
- Form hypotheses: Scientists propose answers to scientific questions, or form **hypotheses,** based on observations they, or other scientists, made.
- Test hypotheses: Scientists devise methods of observing and experimenting to test their predictions.
- Analyze data: Scientists use statistics to analyze data. This analysis tells a scientist whether a hypothesis is supported or not supported by the data.
- Evaluate results: Scientists evaluate both their own results and the results from other scientists.

Scientists use experiments to test hypotheses. A scientific **experiment** uses tightly controlled conditions to test a possible cause-and-effect relationship between variables. In an experiment, there are constants and two types of variables: the independent variable and dependent variables.

- Independent variables: An **independent variable** is a condition that is manipulated by a scientist to determine its effect on a dependent variable. An independent variable is the "cause."
- Dependent variables: A **dependent variable** is measured by a scientist to study the effect of the independent variable. It is the "effect" and depends on the independent variable.
- Constants: Factors that are controlled so that they do not change are **constants.**

A hypothesis is a proposed explanation for a single observation. A scientific **theory,** however, is a proposed explanation for a wide range of observations and experimental results, and is supported by a wide range of evidence.

1. How do scientists use hypotheses? ______________________________

____

2. What is the difference between independent variables and dependent variables? ______________________________

____

____

3. What is a theory? ______________________________

____

Name Period Date

SECTION **1.4** | BIOLOGISTS' TOOLS AND TECHNOLOGY

Study Guide

KEY CONCEPT

Technology continually changes the way biologists work.

VOCABULARY	
microscope	molecular genetics
gene	genomics

MAIN IDEA: Imaging technologies provide new views of life.

Compare and contrast the different types of microscopes and medical imaging techniques.

Type of Technology	Characteristics
Light microscope (LM)	**1.**
Scanning electron microscope (SEM)	**2.**
Transmission electron microscope (TEM)	**3.**
X-ray	**4.**
Magnetic resonance imaging (MRI)	**5.**

MAIN IDEA: Complex systems are modeled on computers.

6. What is a model?

7. Why might scientists use computer models in research instead of conducting an experiment on the real system they would like to study?

MAIN IDEA: The tools of molecular genetics give rise to new biological studies.

8. What is a gene?

9. How are computers used in genomics?

10. How does a gene differ from a genome?

Vocabulary Check

11. The term *genomics* is related to the term *genome*. How does the definition of genome give you a clue about what genomics means?

12. The term *molecular genetics* is made up of two words: *molecular* and *genetics*. What are the meanings of these two words, and how can these words help you to remember what molecular genetics is?

Name Period Date

SECTION 1.4 | BIOLOGISTS' TOOLS AND TECHNOLOGY

Power Notes

Tool or Technology	Details
Light microscope (LM)	**1.**
Scanning electron microscope (SEM)	**2.**
Transmission electron microscope (TEM)	**3.**
X-ray imaging	**4.**
Magnetic resonance imaging (MRI)	**5.**
Functional magnetic resonance imaging (fMRI)	**6.**
Computer modeling	**7.**
Molecular genetics	**8.**

Name Period Date

SECTION 1.4 | BIOLOGISTS' TOOLS AND TECHNOLOGY
Reinforcement

KEY CONCEPT Technology continually changes the way biologists work.

The invention of the microscope in the late 1600s played a major role in the development of biology as a science. Because **microscopes** could greatly magnify very tiny things, scientists discovered microscopic organisms, or microorganisms, and recognized the cell as the basic unit of life. Light microscopes are still used by many biologists, and they are still based on the same principles. Other types of microscopes, called electron microscopes, are also commonly used today.

- Light microscope (LM): Light microscopes use light and glass lenses to magnify objects. Both living and preserved specimens can be observed with light microscopes. These microscopes can clearly magnify specimens about 1500 times.
- Scanning electron microscope (SEM): Scanning electron microscopes use electrons to magnify objects. Electrons that bounce off a specimen are detected and interpreted by a computer to form a three-dimensional image of the specimen. An SEM cannot be used to study a living specimen, but it can clearly magnify a specimen more than 100,000 times.
- Transmission electron microscope (TEM): Transmission electron microscopes are similar to SEMs. They use electrons to magnify specimens, they cannot be used to study living specimens, and they can clearly magnify specimens more than 100,000 times. Unlike SEMs, TEMs use electrons that pass through a specimen. The images are two-dimensional.

Other computer-based technologies are very important in many areas of biology. Computer models are used to study things that would otherwise be impossible to study directly, such as the possible effects of medicines and the potential spread of diseases. Computers are used in genetics research. They can be used to study a single gene, which is a stretch of DNA that stores genetic information, or they can be used to study an organism's genome, which is all of its DNA.

1. What are the differences between light microscopes and electron microscopes?

____

____

____

2. When are computer models most useful?

____

3. What is a gene? What is a genome?

____

____

Name Period Date

SECTION **1.5** BIOLOGY AND YOUR FUTURE

Study Guide

KEY CONCEPT

Understanding biology can help you make informed decisions.

VOCABULARY
biotechnology
transgenic

MAIN IDEA: Your health and the health of the environment depend on your knowledge of biology.

1. Briefly describe three ways in which biology can help you make informed decisions about your health.

____

____

2. Briefly describe why biology and scientific thinking can help you make informed decisions related to the world around you.

____

____

MAIN IDEA: Biotechnology offers great promise but also raises many issues.

3. What is biotechnology?

____

Use the chart below to list the benefits of biotechnology, as well as the risks and ethical concerns about biotechnology.

Benefits	Risks and Ethical Concerns
4.	5.

MAIN IDEA: Biology presents many unanswered questions.

6. Most of our knowledge about DNA was discovered during the past ______________________ years.

7. Many questions go ______________________ because we don't know enough about biology to even come up with those questions.

8. Before the invention of the microscope, people did not know about cells and bacteria. With this in mind, why do you think many questions go unanswered and unasked?

____

____

Vocabulary Check

Each of the vocabulary words has been divided into its root words. Define the roots. Then use the definitions to define the vocabulary word.

9. *Transgenic organism* can be divided into *trans-* and *genic.*

____

____

10. *Biotechnology* can be divided into *bio-* and *technology.*

____

____

Any Questions?

11. What questions do you have about biology or scientific research? List three topics in biology that you want to learn more about, and why they interest you.

____

____

____

____

____

____

Name Period Date

SECTION **1.5** | BIOLOGY AND YOUR FUTURE

Power Notes

Personal health and lifestyle choices:

Environment and society:

Biology

Biotechnology		
Benefits	Risks	Ethical Considerations

Name Period Date

SECTION 1.5 | BIOLOGY AND YOUR FUTURE

Reinforcement

KEY CONCEPT Understanding biology can help you make informed decisions.

An understanding of biology is important for understanding many issues, both on a personal level and on a societal level. Choices that you make about your personal lifestyle and about the environment around you are often related to biology in some way. Your knowledge of biology can help you make informed decisions.

Your health can depend on your knowledge of biology. Knowing the nutrients you need, the causes of diseases, and the things you should avoid, such as cigarettes, can help you lead a longer, healthier life. For example, obesity has been linked to an increased risk of diabetes, stroke, heart disease, and various types of cancer. Similarly, cigarettes have been linked to an increased risk of stroke, heart disease, and cancer. The choices you make affect your health.

Biology can also help you address issues that affect the world around you.

- The environment: The environment around you affects you, and you affect the environment. Biologists study environmental issues such as pollution, biodiversity, habitat preservation, land conservation, and natural resource use, but they don't make decisions about these issues. Instead, all people can use scientific research to help form their own opinions and make their own choices.
- Biotechnology: As we have learned more about living things and how they function, many advances in biotechnology have taken place. **Biotechnology** is the use and application of living things and biological processes. Biotechnology presents both great potential benefits and risks, which you may have to consider throughout your lifetime. Biotechnology is very useful in developing medical treatments and in improving agriculture. However, the long-term effects of many biotechnology applications are unknown.

1. How can a knowledge of biology help you make informed decisions about your health?

____

____

2. Why is an understanding of the environment important?

____

____

3. What is biotechnology?

____

____

CHAPTER 1 | TYPES OF DATA: QUALITATIVE AND QUANTITATIVE

Data Analysis Practice

Two different types of data can be collected during scientific investigations: qualitative data and quantitative data.

The following data were collected by scientists studying the feeding habits of the purple-throated carib hummingbird, *Eulampis jugularis*. This hummingbird is found on several Caribbean islands including Dominica and St. Lucia. It feeds on both nectar and insects. In this study, scientists observed the birds in two different locations on the island of Dominica. One area was a landscaped garden with red ginger and several species of *Heliconia*. The other was a forested area dominated by *Heliconia* (spp.). A summary of the data collected is listed below.

- The males of the species are larger than the females.
- The males' bills are shorter and less curved than those of the female.
- The hummingbirds in the landscaped garden spent 21% of the time feeding. Of that time, 20% was spent feeding on insects and 80% was spent feeding on nectar.
- The males get nectar from a species of *Heliconia* in which the flowers have shorter and less curved petals.
- The females get nectar from a species of *Heliconia* in which the flowers have longer and more curved petals.
- The hummingbirds in the forested area spent 23% of the time feeding. Of that time, 92% was spent feeding on insects and 8% was spent feeding on nectar.

1. **Identify** Which are the quantitative data in the example above? Explain.

____

____

____

2. **Identify** Which are the qualitative data in the example above? Explain.

____

____

____

3. **Synthesize** For each of the qualitative observations, give an example of how it could be transformed to quantitative data.

____

____

____

Name Period Date

CHAPTER 1 | EVOLUTION AND NATURAL SELECTION
Pre-AP* Activity

Chapter 1 introduces evolution as a process of species adaptation. It is important to understand that evolution is often a relatively slow process. Most adaptations arise over the course of many generations, not suddenly. The process by which this occurs is called natural selection.

MUTATION

DNA is the molecule that stores genetic information in organisms. A change in the DNA of an organism is called a mutation. Some mutations cause an organism to bear a trait that is new or different from the traits of its predecessors. Were it not for mutation, life as we know it—with its millions of species and the variety and diversity we see on Earth today—could not have evolved. If the effect of a mutation helps an organism survive and thus reproduce in its environment, we call this effect an adaptation. It usually takes a long time for adaptations to evolve, and even longer for an adaptation to become common or prevalent in a population.

The cicada's basic shape resembles many other flying insect species. The thorn bug, a close relative of the thorn mimic tree hopper on page 10 of your book, has a distinct shape that allows this insect to blend in with real thorns on the stems of plants.

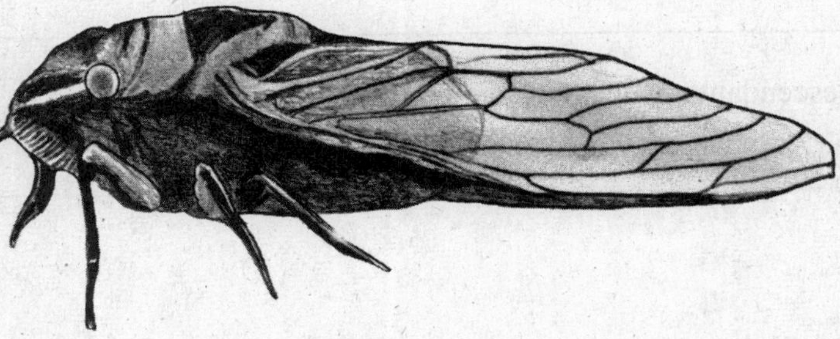

ADAPTATION AND NATURAL SELECTION

When it comes to nature and evolution, we tend to focus our attention on the most striking examples of adaptation, such as the thorn bug's distinctly thornlike shape. Keep in mind that the cicada is just as well adapted to its environment as the thorn bug is to its environment. How did natural selection lead to the evolution of these two bugs?

Imagine that, long ago, the common ancestor of the thorn bug and the cicada was a small flying insect similar in shape to the cicada. A mutation in the DNA of one of these ancestral insects resulted in a slight extension on its exoskeleton just above and behind the head. Because the extension was in no way detrimental to the insect, the insect reproduced, passing on this trait to its offspring. Now imagine that the insects moved into an environment where thorny trees and bushes were prevalent. Birds preyed upon the insects as they fed on the sap of the plants.

1. With mutations occurring and accumulating over generations, birds preying on insects, and surviving insects reproducing, describe how the modern thorn bug might have evolved.

2. In what form is a trait passed on from parent to offspring, and, in effect, what is natural selection "selecting" for?

3. In what sense is natural selection a process of elimination?

4. Draw two examples of intermediate descendants that show how a cicada-like insect might have evolved into a thorn bug.

Name Period Date

CHAPTER 1 | OPTICS OF A MICROSCOPE

Pre-AP Activity

You have learned in Chapter 1 that the simple light microscope is still one of the most important tools in biology. In this activity you will learn about the optics of a microscope, and how the interplay of its lenses allows us to see the microscopic aspects of our world.

FORMING AN IMAGE

For you to see, light must be reflected off an object and bent, or refracted, through the lenses in your eye. These lenses, like those in a microscope, are converging lenses. Light is refracted to the focal point (*F*), where all the rays meet. The distance from the center of a lens to its focal point is the focal length (*f*). A converging lens has two focal points (*F* and *F′*) because it is curved equally on both sides. Where an object is in relation to the focal point determines the size and orientation of the image produced.

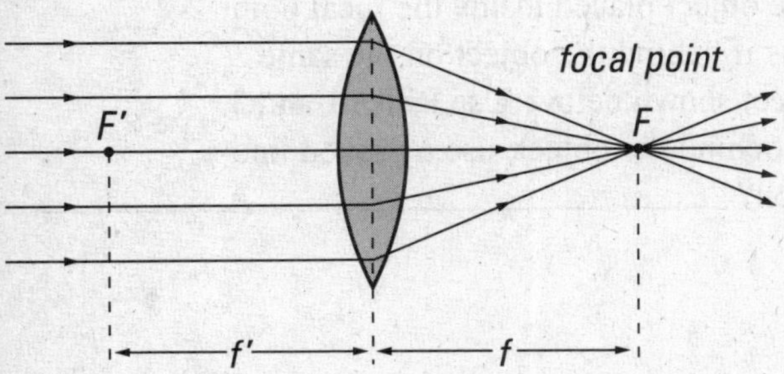

RAY DIAGRAMS

A ray diagram is used to trace the light rays that form an image. Three rules apply. We will use the special case of an object located 2*f* from a lens to explain the rules. As is customary, we will show the rays originating from a single point at the top.

Rule 1 A ray reflected from the object, traveling parallel to the axis, is refracted through the focal point *F* on the far side of the lens.

Rule 2 A ray reflected through the focal point *F′* on the near side of the lens is refracted parallel to the axis.

Rule 3 A ray reflected through the lens center goes straight through without refraction.

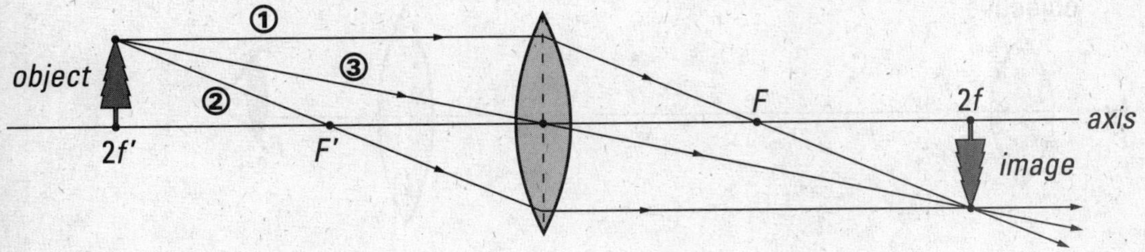

For an object placed 2*f* from the center of a lens, the image formed is the same size as the object. An image that forms directly from the refracted rays is the real image.

1. The example just given shows an object that is placed outside the focal point of a lens. What is the orientation of the image formed? ______________________

2. Draw a ray diagram for the object shown in the diagram below, first placed at point A and then point B. Only two rays are needed to create the image; use Rules 1 and 2. Which image is larger and how do you account for the difference?

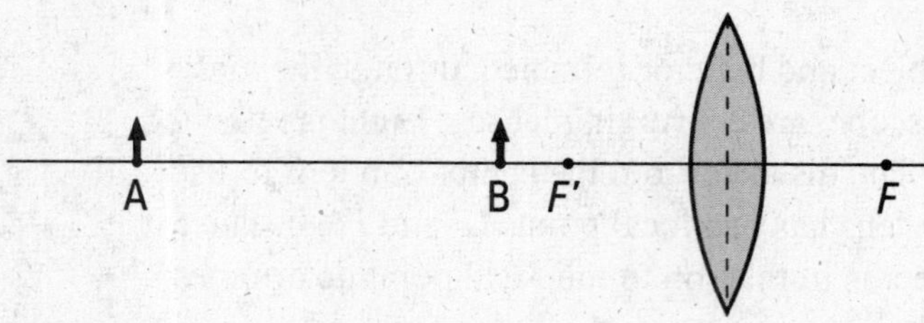

3. A magnifying glass uses a converging lens, with the object placed inside the focal point. The image created is a virtual image that appears as if behind the object on the same side of the lens. Draw the virtual image for the object shown below. Use Rules 1 and 3 to trace the rays back to the point where they meet behind the object; use a dashed line to represent the virtual image. What is its orientation? ______________________

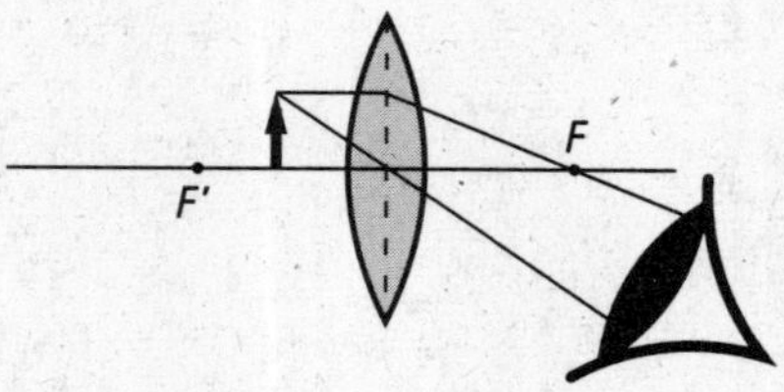

4. Draw the ray diagram for the compound microscope shown below. A compound microscope uses two lenses placed in a connecting barrel. Each produces an image: an objective image and an eyepiece image. Why is the object being magnified placed behind the focal point of the objective lens and not in front, as with the magnifying glass? What is the orientation of each image? Is the final image real or virtual? (Hint: The image of one lens can serve as the object of another.)

Name Period Date

CHAPTER 1 | BIOLOGY IN THE 21ST CENTURY

Vocabulary Practice

biosphere	ecosystem	dependent variable
biodiversity	homeostasis	constant
species	evolution	theory
biology	adaptation	microscope
organism	observation	gene
cell	data	molecular genetics
metabolism	hypothesis	genomics
DNA	experiment	biotechnology
system	independent variable	transgenic

A. Categorize Words For the terms below, write **L** next to words that can describe living things. Write **T** next to words that can describe technology. Write **B** next to words that can describe both.

1. _____ organism _____ species _____ biotechnology

_____ cell _____ transgenic _____ molecular genetics

For the terms below, write **E** next to words that can describe the external environment of living things. Write **I** next to words that can describe the internal environment of living things. Write **B** next to words that can describe both.

2. _____ homeostasis _____ system _____ ecosystem

_____ biosphere _____ metabolism _____ biology

For the terms below, write **G** next to words that are related to groups of living things. Write **I** next to words that describe individual living things. Write **B** next to words that can describe both.

3. _____ biodiversity _____ cell _____ biosphere

_____ organism _____ adaptation _____ evolution

Name Period Date

VOCABULARY PRACTICE, CONTINUED

B. Vector Vocabulary Define the words in the boxes. On the line across each arrow, write a phrase that describes how the words in the boxes are related to each other.

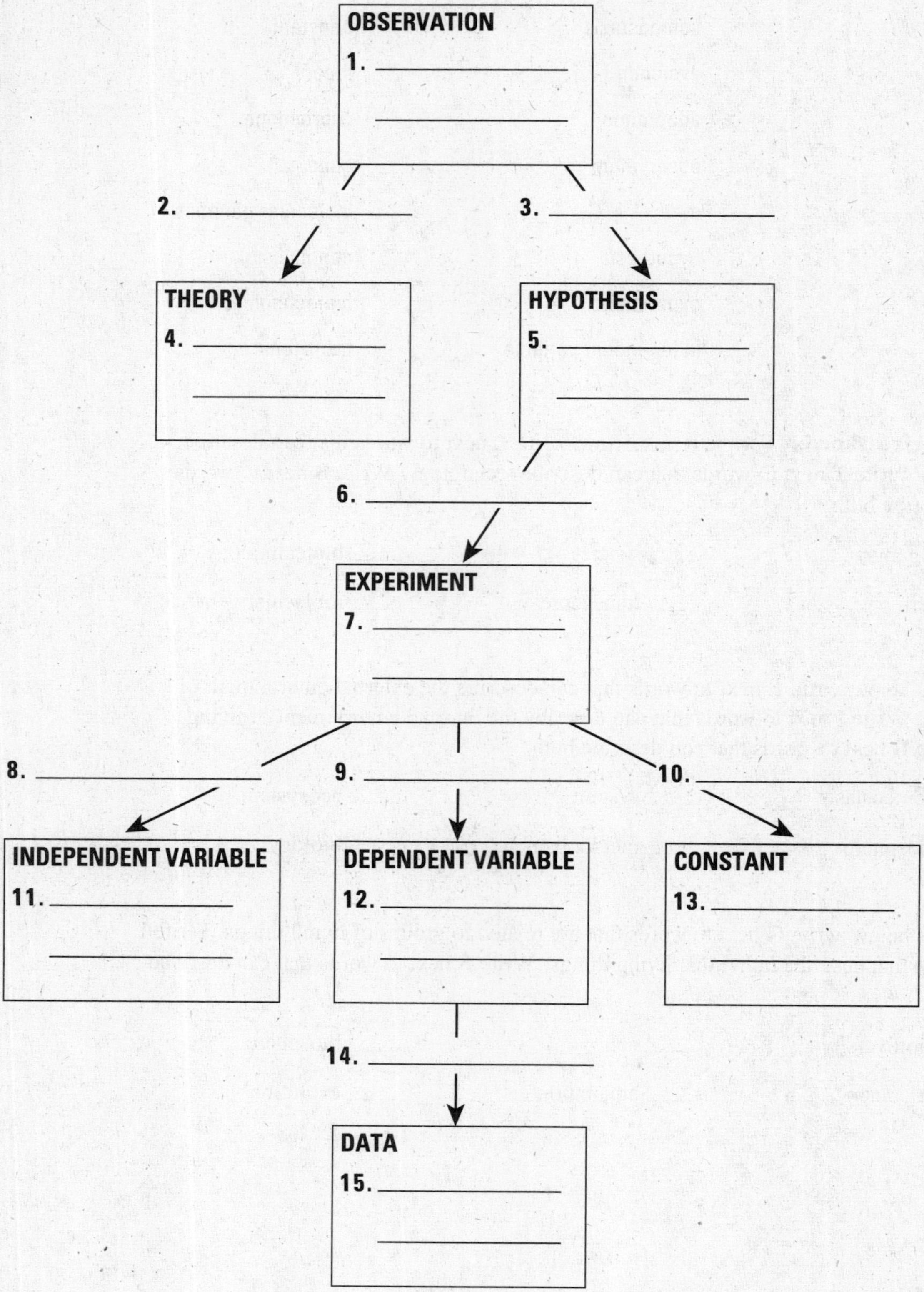

Name Period Date

VOCABULARY PRACTICE, CONTINUED

C. Stepped-Out Vocabulary Define each word. Then write two additional facts that are related to the word.

WORD	DEFINITION	MORE INFORMATION
Example metabolism	all chemical processes that build up or break down materials in living things	chemical energy is needed
		animals eat other organisms to get their chemical energy
1. biology		
2. microscope		
3. evolution		
4. adaptation		
5. DNA		
6. gene		
7. genomics		

Name Period Date

D. Words in Context Answer the questions to show your understanding of the vocabulary words.

1. If I use a **microscope,** do I see things too small to be seen or things too far away to be seen?

____

2. Is **homeostasis** the maintenance of constant conditions or all of the chemical processes that build up and break down materials?

____

3. Which is the **independent variable** in an experiment, the cause or the effect?

____

4. Is a **gene** all of an organism's DNA or only a segment of DNA?

____

5. Is an **adaptation** in biology made by choice or is it inherited?

____

6. Would a **constant** be manipulated or kept the same in an experiment?

____

7. Does a **theory** answer one scientific question or does it explain many observations?

____

8. When I am in an **ecosystem,** do I interact with living things, nonliving things, or both living and nonliving things?

____

9. Where would more **biodiversity** be found, near Earth's equator or near Earth's poles?

____

10. Which is a **species,** a group of parts that interact to form a whole or a type of living things that can reproduce by interbreeding?

____

11. Does a **transgenic** organism travel a lot or does it have genes from a different type of living thing?

____

12. If I form a **hypothesis,** do I propose an answer to a question or do I use negative feedback?

____

Name Period Date

SECTION
2.1
ATOMS, IONS, AND MOLECULES
Study Guide

KEY CONCEPT
All living things are based on atoms and their interactions.

VOCABULARY		
atom	ion	molecule
element	ionic bond	
compound	covalent bond	

MAIN IDEA: Living things consist of atoms of different elements.

1. How are atoms and elements related?

____

2. Sketch the structure of an atom. Label the protons, neutrons, nucleus, and electrons.

3. How do compounds differ from elements?

____

MAIN IDEA: Ions form when atoms gain or lose electrons.

4. What is an ion?

____

5. Why does an ion have an electrical charge?

____

____

6. In the spaces provided below, sketch how both positive and negative ions form. Label the nucleus and the electrons. Use Figure 2.3 as a reference.

MAIN IDEA: Atoms share pairs of electrons in covalent bonds.

7. What is a covalent bond?

____

8. What determines the number of covalent bonds that an atom can form?

____

Vocabulary Check

element	compound	ion	molecule

______________ **9.** atoms held together by covalent bonds

______________ **10.** composed of different types of atoms

______________ **11.** composed of one type of atom

______________ **12.** atom that has gained or lost electrons

13. What is the difference between how ionic and covalent bonds form?

____

____

Name Period Date

SECTION **2.1** | ATOMS, IONS, AND MOLECULES

Power Notes

Atom:

Energy levels:

Outermost energy levels:

Element:

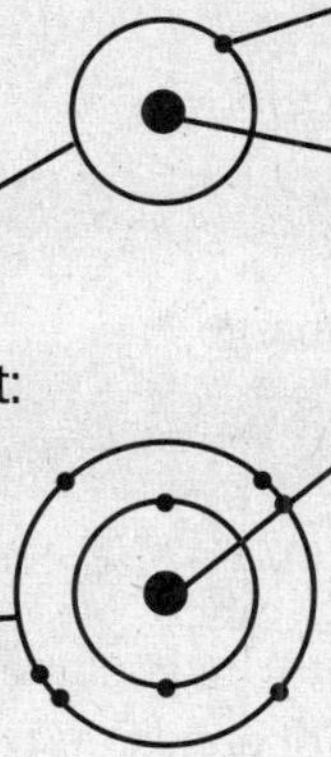

Element:

1. ______________

2. ______________

Composed of:

3. ______________

4. ______________

Compound:

Ionic bonds:

Ions:

Positive ions:

Negative ions:

Covalent bonds:

Molecules:

Name Period Date

SECTION **2.1** ATOMS, IONS, AND MOLECULES
Reinforcement

KEY CONCEPT All living things are based on atoms and their interactions.

All matter, whether living or nonliving, is made of the same tiny building blocks, called atoms. An **atom** is the smallest basic unit of matter. All atoms have the same basic structure, composed of three smaller particles.

- Protons: A proton is a positively charged particle in an atom's nucleus. The nucleus is the dense center of an atom.
- Neutrons: A neutron has no electrical charge, has about the same mass as a proton, and is also found in an atom's nucleus.
- Electrons: An electron is a negatively charged particle found outside the nucleus. Electrons are much smaller than either protons or neutrons.

Different types of atoms are called **elements,** which cannot be broken down by ordinary chemical means. Which element an atom is depends on the number of protons in the atom's nucleus. For example, all hydrogen atoms have one proton, and all oxygen atoms have 16 protons. Only about 25 different elements are found in organisms. Atoms of different elements can link, or bond, together to form **compounds.** Atoms form bonds in two ways.

- Ionic bonds: An **ion** is an atom that has gained or lost one or more electrons. Some atoms form positive ions, which happens when an atom loses electrons. Other atoms form negative ions, which happens when an atom gains electrons. An **ionic bond** forms through the electrical force between oppositely charged ions.
- Covalent bonds: A **covalent bond** forms when atoms share one or more pairs of electrons. A **molecule** is two or more atoms that are held together by covalent bonds.

1. What are the parts of an atom?

____

2. What makes atoms of one element different from the atoms of another element?

____

3. How are ionic bonds and covalent bonds different?

____

____

____

Name Period Date

SECTION
2.2

PROPERTIES OF WATER
Study Guide

KEY CONCEPT

Water's unique properties allow life to exist on Earth.

VOCABULARY		
hydrogen bond	solution	acid
cohesion	solvent	base
adhesion	solute	pH

MAIN IDEA: Life depends on hydrogen bonds in water.

1. What is a polar molecule?

____

2. Explain why water is a polar molecule.

____

____

3. What is a hydrogen bond?

____

4. Describe where a hydrogen bond can form among water molecules.

____

Complete the table by writing short descriptions about the properties of water.

Property	Description
High specific heat	**5.**
Cohesion	**6.**
Adhesion	**7.**

CHAPTER 2
Chemistry of Life

MAIN IDEA: Many compounds dissolve in water.

8. What is the difference between a solvent and a solute?

9. What types of substances dissolve easily in water?

10. What types of substances do not dissolve easily in water?

MAIN IDEA: Some compounds form acids or bases.

11. Take notes about the characteristics of acids and bases in the table below.

Characteristic	Acid	Base
Effect on H^+ concentration in a solution		
Effect on pH		

Vocabulary Check

12. In the space below, sketch a solution using the Visual Vocab on page 42 as a reference. Label the solution, solvent, and solute. Next to these labels, write brief definitions for the terms.

Name Period Date

SECTION **2.2** | PROPERTIES OF WATER

Power Notes

Polar molecules:

Nonpolar molecules:

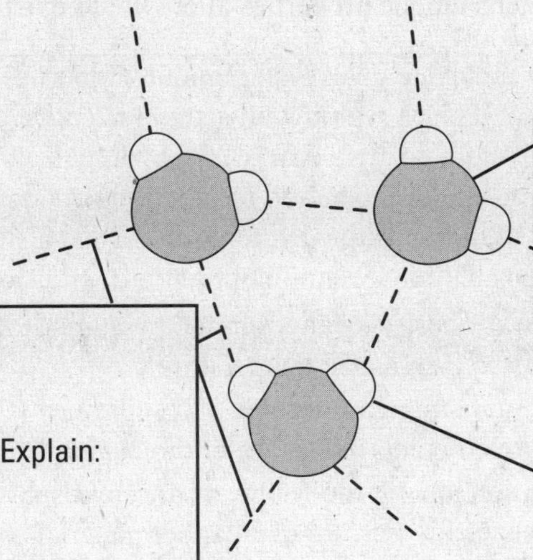

1.

2. Explain:

Atom:

Charge:

Atom:

Charge:

Properties of water related to hydrogen bonds:

1.

2.

3.

Solutions:

Solvents:

Solutes:

Acids:

pH:

Neutral:

pH:

Bases:

pH:

Name Period Date

SECTION 2.2 | PROPERTIES OF WATER

Reinforcement

KEY CONCEPT Water's unique properties allow life to exist on Earth.

The structure of the water molecule gives water unique properties. Water is a polar molecule, which means that it has a region with a slight negative charge (the oxygen atom), and a region with a slight positive charge (the hydrogen atoms). The oppositely charged regions of water molecules interact to form hydrogen bonds. A **hydrogen bond** is an attraction between a slightly positive hydrogen atom and a slightly negative atom. Hydrogen bonds are responsible for several important properties of water.

- High specific heat: Water resists changes in temperature; it must absorb a large amount of heat energy to increase in temperature.
- Cohesion: The attraction among molecules of a substance is called **cohesion.** Cohesion due to hydrogen bonds makes water molecules "stick" together.
- Adhesion: The attraction among molecules of different substances is called **adhesion.** Water molecules "stick" to many other materials because of hydrogen bonds.

Many compounds that are important for life dissolve in water. Water is the largest component of cells' interiors, and chemical reactions in the cell take place in this water. When one substance dissolves in another, a **solution** is formed. The substance present in the greatest amount is called the **solvent.** Substances that are present in lower amounts and dissolve in the solvent are called **solutes.** Polar solvents, such as water, dissolve polar molecules and ions.

When some substances dissolve in water they break up into ions. A compound that releases a hydrogen ion (a proton) when it dissolves in water is an **acid. Bases** are compounds that remove, or accept, hydrogen ions. A solution's acidity, or its hydrogen ion concentration, is measured on the pH scale. An acid has a low pH (pH below 7) and a high hydrogen ion concentration. A base has a high pH (pH above 7) and a low hydrogen ion concentration. Organisms must maintain a stable pH. Even a small change in pH can disrupt many biological processes.

1. How do hydrogen bonds form?

____

____

____

2. What are the two parts of a solution?

____

3. What does pH measure?

____

Name Period Date

SECTION
2.3
CARBON-BASED MOLECULES
Study Guide

KEY CONCEPT
Carbon-based molecules are the foundation of life.

VOCABULARY		
monomer	lipid	amino acid
polymer	fatty acid	nucleic acid
carbohydrate	protein	

MAIN IDEA: Carbon atoms have unique bonding properties.

1. Why is carbon often called the building block of life?

____

____

2. What ability allows carbon atoms to form a large number of molecules?

____

____

3. In the space below, sketch the three basic structures of carbon-based molecules: straight chain, branched chain, and ring.

CHAPTER 2
Chemistry of Life

MAIN IDEA: Four main types of carbon-based molecules are found in living things.
Complete the table with functions and examples of each type of carbon-based molecule.

Molecule Type	Functions	Examples
Carbohydrate	**4.**	**5.**
Lipid	**6.**	**7.**
Protein	**8.**	**9.**
Nucleic acid	**10.**	**11.**

12. What determines a protein's structure and function?

____

13. What are nucleic acids made of?

____

Vocabulary Check

14. The prefix *mono-* means "one," and the prefix *poly-* means "many." How are these meanings related to the terms *monomer* and *polymer*?

____

____

____

Name Period Date

SECTION **2.3** | CARBON-BASED MOLECULES

Power Notes

Monomer:	→	Polymer:

Type of Molecule	Functions	Example
1. Carbohydrate		Polymer – cellulose 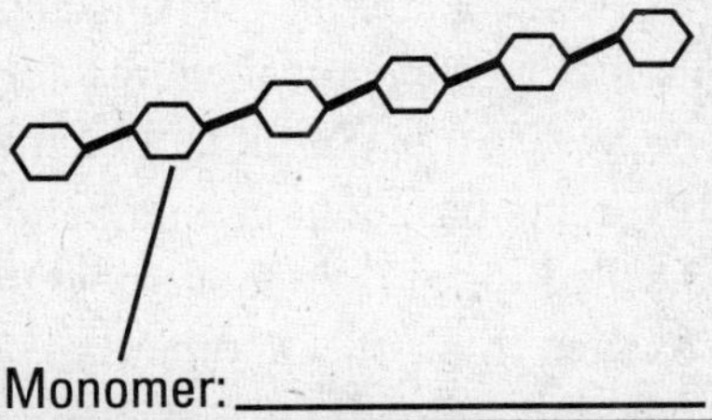 Monomer: ________
2.		Molecule: ________  ________ ________
3.		Polymer: ________ 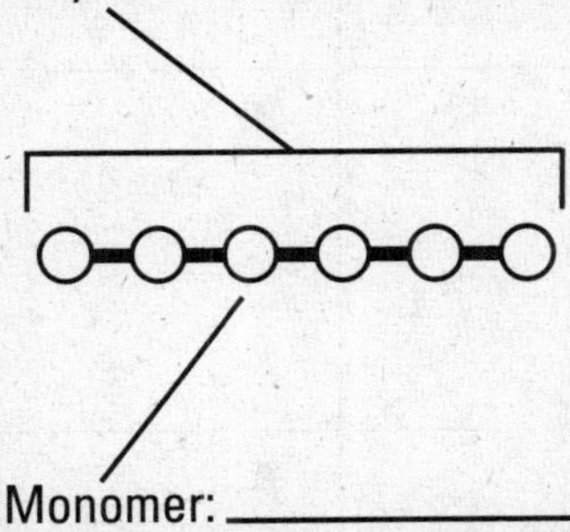 Monomer: ________

Nucleic Acids	Types:	Functions:

Name Period Date

SECTION
2.3

CARBON-BASED MOLECULES

Reinforcement

KEY CONCEPT Carbon-based molecules are the foundation of life.

Carbon atoms are the basis of most molecules that make up living things. Many carbon-based molecules are large molecules called polymers that are made of many smaller, repeating molecules called monomers. There are four main types of carbon-based molecules in living things.

- **Carbohydrates** include sugars and starches, and are often broken down as a source of chemical energy for cells. Some carbohydrates are part of cell structure, such as cellulose, which makes up plant cell walls.
- **Lipids** include fats and oils and, like carbohydrates, are often broken down as a source of chemical energy for cells. One type of lipid, called a phospholipid, makes up most of all cell membranes.
- **Proteins** have a large number of structures and functions. Some proteins are needed for muscle movement; another protein, called hemoglobin, transports oxygen in blood. Another type of proteins, called enzymes, speed up chemical reactions in cells.
- **Nucleic acids** are molecules that store genetic information and build proteins. DNA stores genetic information in cells, and RNA helps to build the proteins for which DNA codes.

Type of Molecule	Functions	Example
Carbohydrate	**1.**	**2.**
Lipid	**3.**	**4.**
Protein	**5.**	**6.**
Nucleic acid	**7.**	**8.**

Name Period Date

SECTION **2.4** | CHEMICAL REACTIONS

Study Guide

KEY CONCEPT

Life depends on chemical reactions.

VOCABULARY		
chemical reaction	bond energy	exothermic
reactant	equilibrium	endothermic
product	activation energy	

MAIN IDEA: Bonds break and form during chemical reactions.

1. Label the reactants and products in the chemical reaction shown below. Write brief definitions for these terms next to their labels.

$$CH_4 + 2O_2 \longrightarrow CO_2 + 2H_2O$$

2. What causes chemical bonds to break during a reaction?

3. What is bond energy?

4. In a chemical equation, what symbol is used to show that a chemical reaction goes in both directions?

5. When does a chemical reaction reach equilibrium?

MAIN IDEA: Chemical reactions release or absorb energy.

6. The ______________________ of the reactants and products determines whether energy will be released or absorbed during a chemical reaction.

7. Before a chemical reaction can start, ______________________ must be absorbed by the reactants. The amount that must be absorbed to start the reaction is called the ______________________ .

8. In an exothermic reaction, the products have a ______________________ bond energy than the reactants. Overall, energy is ______________________ .

9. In an endothermic reaction, the products have a ______________________ bond energy than the reactants. Overall, energy is ______________________ .

Vocabulary Check

10. Write one sentence that uses the words *chemical reaction, reactant,* and *product*.

____

11. Write your own analogy to remember the meaning of *activation energy*.

____

____

12. The term *equilibrium* is based on two Latin roots that mean "equal" and "balance." How do these meanings tell you the meaning of *equilibrium* in a chemical reaction?

____

____

13. The prefix *exo-* means "out," and the prefix *endo-* means "in." What do these prefixes tell you about *exothermic* and *endothermic* reactions?

____

____

SECTION **2.4** | CHEMICAL REACTIONS
Power Notes

Chemical Reaction

$6O_2 + C_6H_{12}O_6$ → $6CO_2 + 6H_2O$

1.

2.

3.

Bond energy:

Chemical equilibrium:

Exothermic:

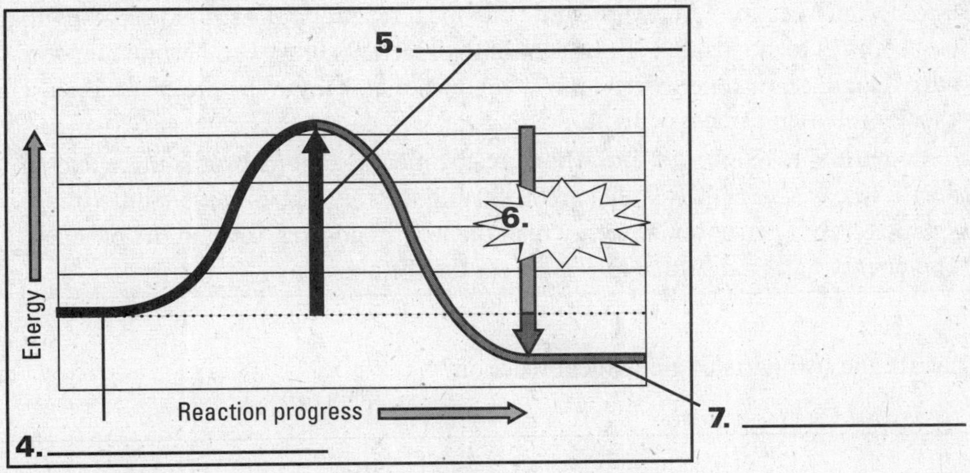

Endothermic:

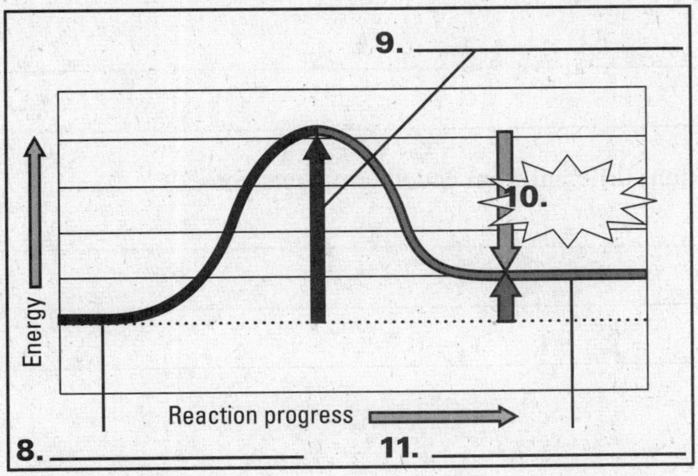

Name Period Date

SECTION **2.4** CHEMICAL REACTIONS

Reinforcement

KEY CONCEPT Life depends on chemical reactions.

At the most fundamental level, every process that takes place in an organism depends on chemical reactions. In a **chemical reaction,** substances are changed into different substances by the breaking and forming of chemical bonds. The substances that are present at the start of a chemical reaction, and are changed by the reaction, are called **reactants.** The substances that are formed by a chemical reaction are the **products.**

Chemical bonds must be broken in the reactants and new ones must be formed in the products. Energy must be added to break chemical bonds. In contrast, energy is always released when new bonds form. The amount of energy needed to break a bond, or the amount of energy released when a bond forms, is called **bond energy.**

All chemical reactions require the input of at least a small amount of energy in order for bonds to break in the reactants and for the reaction to start. The energy needed to start a chemical reaction is the **activation energy.** In general, there are two types of energy changes that can occur during a chemical reaction.

- Exothermic reaction: An **exothermic** chemical reaction releases more energy than it absorbs. The bonds that are broken in the reactants of an exothermic reaction have a higher bond energy than the new bonds that form in the products. Energy is usually released as heat or light.
- Endothermic reaction: An **endothermic** chemical reaction absorbs more energy than it releases. The bonds that are broken in the reactants of an endothermic reaction have a lower bond energy than the new bonds that form in the products. The energy that is absorbed makes up for the difference.

1. What are the two parts of a chemical reaction?

____

2. What is activation energy?

____

____

3. How are exothermic reactions different from endothermic reactions?

____

____

____

Name Period Date

SECTION **2.5** | ENZYMES

Study Guide

KEY CONCEPT

Enzymes are catalysts for chemical reactions in living things.

VOCABULARY	
catalyst	substrate
enzyme	

MAIN IDEA: A catalyst lowers activation energy.

1. What is activation energy?

____

____

2. Take notes about catalysts in the chart below. In the first two boxes, write detail notes about the main functions of catalysts. In the third box, write a detail about another characteristic.

A catalyst lowers activation energy.

→

→

→

3. When a catalyst is present, more / less activation energy is needed to start a chemical reaction.

____

MAIN IDEA: Enzymes allow chemical reactions to occur under tightly controlled conditions.

4. Take notes about enzymes by filling in the Main Idea Web below.

Why enzymes are necessary:

How structure affects function:

Enzymes

Important factors in enzyme structure:

Lock-and-key model:

5. How do enzymes weaken the bonds in substrates?

____

____

Vocabulary Check

6. The word *catalyst* comes from the Greek word meaning "to dissolve." How does this definition relate to the meaning of *catalyst*?

____

7. How are substrates like keys and enzymes like locks?

____

____

Name Period Date

SECTION 2.5 | ENZYMES

Power Notes

A Catalyst:

1.

2.

Enzymes:

Homeostasis and enzymes:

Enzyme structure and function:

Substrates and Lock-and-Key Model of Enzyme Function	
Substrates:	Lock-and-key model:

CHAPTER 2
Chemistry of Life

Name Period Date

SECTION
2.5

ENZYMES
Reinforcement

KEY CONCEPT Enzymes are catalysts for chemical reactions in living things.

Chemical reactions require the addition of energy, called activation energy, to take place. Even if a chemical reaction starts, it may not happen very quickly. However, both the activation energy and the rate of a chemical reaction can be changed by a chemical catalyst. A **catalyst** is a substance that decreases the activation energy for a reaction and increases the rate of the reaction.

In living things, **enzymes** are catalysts for chemical reactions. Almost all enzymes are proteins, and almost every process in living things needs enzymes. The function of each enzyme depends on its structure. A change in biological conditions within an organism can affect the shape of an enzyme, which can decrease or prevent an enzyme from working properly. For example, enzymes function best in a small range around an organism's normal temperature and pH.

The shape of an enzyme is important because it allows only certain molecules to bind to the enzyme. The specific molecules that an enzyme acts on are called **substrates.** One way to think of enzyme function is called the lock-and-key model. Substrates bind to an enzyme like the way in which a key fits into a lock. If an enzyme's structure changes, the substrates cannot bind to the enzyme.

- Substrates bind to an enzyme. The enzyme brings molecules close together so that they can react with one another.
- The bonds inside the substrates are stretched slightly out of position, which weakens the bonds. Less energy is needed to break weaker bonds.
- The reaction takes place and the product is released from the enzyme. The enzyme can then bind to more of the substrate molecules.

1. What is a catalyst?

____

____

2. Where are enzymes found?

____

3. How does an enzyme work?

____

____

____

____

Name Period Date

CHAPTER 2 | INDEPENDENT AND DEPENDENT VARIABLES

Data Analysis Practice

During experiments scientists manipulate or change an independent variable to observe or measure its effects on a dependent variable.

Chemists working for a pharmaceutical company carry out an experiment in which they measure the amount of time it takes for different amounts of an antacid to neutralize 1 mL of hydrochloric acid (HCl). The results of their experiment are shown in the graph below.

GRAPH 1: ANTACID AMOUNTS AND NEUTRALIZATION TIMES

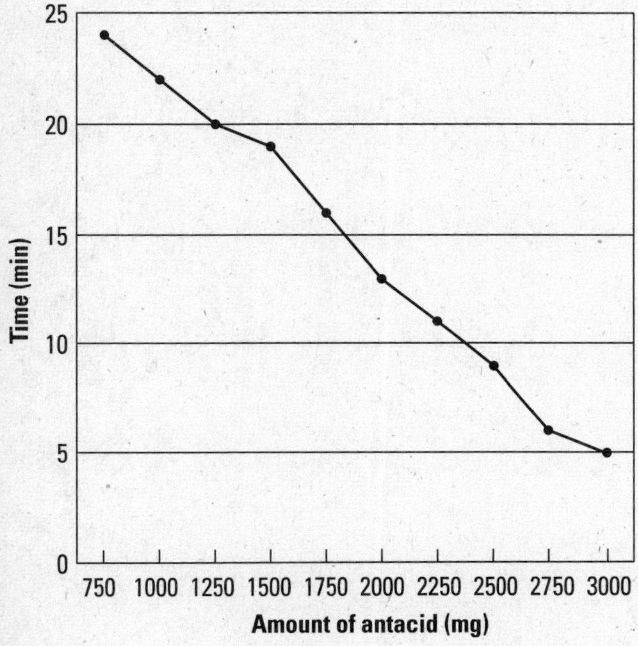

1. Identify What are the independent and dependent variables in this experiment?

____

____

____

2. Analyze What is the relationship between the amount of antacid and the length of time to neutralize the acid?

____

____

Name Period Date

CHAPTER 2

SURFACE TENSION AND COHESION

Pre-AP Activity

In Chapter 2, you have learned that water is a highly polar molecule that forms hydrogen bonds with other water molecules. Many of the physical properties of water—such as surface tension, capillarity, adhesion, and cohesion—result from hydrogen bonding.

INTERMOLECULAR FORCES CAUSE SURFACE TENSION

Many physical properties depend on the intermolecular forces that exist between the particles of the substance. The most important force that exists between water particles is the hydrogen bond. If you could see the forces of hydrogen bonds in a drop of water, you would see that molecules in the interior of the drop are surrounded by and acting upon one another in all directions, and there is no net force on any of them. Molecules at the surface of the drop are subject to the force of hydrogen bonds only from the side and from below. The result of this uneven force is that the water molecules on the surface are pulled toward the center. This net inward force makes the surface contract and act like an elastic skin. Because the intermolecular forces are so strong, it is hard to overcome them and break through the water's surface. In other words, water has a high surface tension.

INTERMOLECULAR FORCES CAUSE CAPILLARITY

When water is placed in a narrow tube, it rises. This property, typical of polar liquids, is called capillarity. Two kinds of forces are responsible for capillarity: cohesive forces and adhesive forces. Cohesive forces are the intermolecular forces that exist between the molecules of a liquid, such as the hydrogen bonds between molecules of water. Adhesive forces are the forces that exist between the liquid molecules and a solid, as in drops of water on a window, or a volume of water in a test tube. Adhesion occurs when the solid has polar bonds. For example, the glass of a test tube has many oxygen atoms with partial negative charges. These charges attract the positive ends of water molecules and cause the water molecules that touch the glass to cling to its surface. Because water also has strong cohesive forces, the water molecules in the interior of the tube are pulled slightly upward by those that are touching the glass. The concave shape of the water's surface in the tube, called the meniscus, shows that the adhesive forces between water molecules and the glass are stronger than the cohesive forces between water molecules.

SURFACTANTS AND SURFACE TENSION

Detergents are substances, known generally as surfactants, which have a polar head and a nonpolar tail. When surfactants interact with water, their polar heads form hydrogen bonds with water molecules, but their tails do not. This behavior reduces the cohesive forces in water because fewer hydrogen bonds are formed. As a result, the overall forces holding the water together are reduced, and the surface tension decreases.

You will investigate the forces that hold a drop of water together. Remember that the molecules on the surface of the drop are pulled toward the interior by unbalanced intermolecular forces. In order to balance this inward force, the molecules at the surface pack together, reducing the surface area and causing the surface of the drop to have a "skin."

Experiment

In this activity, you will count the number of drops of water, rubbing alcohol, and detergent solution that can be placed on a penny without spilling over. By comparing these numbers, you can compare the strengths of the intermolecular forces—and thus the surface tensions—of these substances.

MATERIALS

Detergent solution
(1 part liquid dishwashing detergent: 1 part water)
Dropper
Paper towels (2)
Penny
Rubbing alcohol
Tap water

1. Place a penny heads up on a paper towel on your lab table. Be sure the penny is flat.
2. Make a data table for recording your data. If time allows, plan on doing several trials with each liquid and averaging the data.
3. Fill a dropper with tap water.
4. Holding the dropper close to the penny but not touching it, count the number of drops of water that you can drop onto the penny before they spill over the edge. Record your data in the data table. Rinse the dropper, and dry the penny with a paper towel.
5. Fill the dropper with rubbing alcohol and repeat step 4.
6. Fill the dropper with detergent solution and repeat step 4.

1. On a separate piece of paper, draw a diagram showing the intermolecular forces in the drop of water on the penny.

2. Rank the substances from highest to lowest in terms of surface tension, and list the number of drops of each substance the penny could hold before spilling occurred.

3. Infer why rubbing alcohol has a lower surface tension than water.

4. Compare your results with those of your classmates. What could account for differences in the number of drops of the same liquid that piled up on the pennies?

Name Period Date

CHAPTER 2 | ENZYME ACTION

Pre-AP Activity

In Chapter 2, you have learned about carbon–based molecules, chemical reactions, enzymes and the relationships between them. Enzymes are necessary for every process in an organism. More specifically, enzymes are an integral part of the digestive system, where they act as catalysts to break down carbon–based molecules through a series of chemical reactions.

CARBON–BASED MOLECULES

Carbohydrates, lipids, and proteins are three examples of carbon–based molecules that can be broken down through enzymatic action.

- Carbohydrates, including sugars and starches, are used by the body for energy. Sugars can be classified as monosaccharides, disaccharides, or polysaccharides. Monosaccharides are simple sugars, disaccharides are molecules of two simple sugars bonded together, and polysaccharides are polymers made of monosaccharides. Starches, glycogen, and cellulose are all examples of polysaccharides.
- Lipids are needed for energy, to supply structural components of cell membranes and myelin sheaths, and for the production of some hormones. The most common type of lipids are fats and oils. Fats and oils are made of a glycerol molecule bonded to a fatty acid molecule. When three fatty acids are bonded to a glycerol, they form a triglyceride. Fatty acids that are joined by carbon-to-carbon single bonds are known as saturated fats and those joined by at least one carbon-to-carbon double bond are unsaturated fats.
- Proteins, the raw materials needed for cell and tissue repair and growth, are polymers made of monomers called amino acids. Amino acids are bonded together by a specific type of covalent bond, a peptide bond. When a chain of amino acids bond together, they form a polypeptide. A protein is made of one or more polypeptides. Most enzymes are proteins.

DIGESTION AND ENZYMES

Most carbon–based molecules do not provide any nutritional benefit until they are broken down into simpler substances through digestion. In the human body, the majority of digestion takes place in three locations: the mouth, the stomach, and the small intestine. Digestion begins in the mouth and finishes in the small intestine. The pancreas also plays an important role in digestion. Pancreatic secretions aid digestion in the small intestine. Enzymes released in the mouth, stomach, and small intestine facilitate the digestion of specific carbon–based molecules by breaking them down into simpler substances.

Most enzymes are sensitive to their environment and require specific conditions to function optimally. One such condition is pH, which is a measure of how acidic something is. The optimal pH value for most enzymes is between 6 and 8. Enzymatic action can decrease if the pH becomes more acidic or more basic.

Name Period Date

Human Digestive Organ	Enzyme	Function
Mouth (pH 7.0)	Salivary amylase	Breaks starches (polysaccharides) down into disaccharides and trisaccharides
	Lingual lipase	Breaks triglycerides (fats and oils) and other lipids into fatty acids and diglycerides
Stomach (pH 1–3)	Pepsin	Breaks proteins down into peptides
	Gastric lipase	Breaks short–chain triglycerides (fats and oils) in fat molecules in milk into fatty acids and monoglycerides
Pancreas (secretes enzymes into the stomach)	Pancreatic amylase	Breaks starches (polysaccharides) into maltose (disaccharide) and maltotriose (trisaccharide)
	Trypsin	Breaks proteins into peptides
	Pancreatic lipase	Breaks emulsified triglycerides (fats and oils) into fatty acids and monoglycerides
Small Intestine (pH 6–6.5)	Maltase	Breaks maltose into glucose
	Sucrase	Breaks sucrose into glucose and fructose
	Lactase	Breaks lactose into glucose and galactose
	Dipeptidase	Breaks dipeptides into amino acids

Answer the following questions on a separate piece of paper.

1. What inferences can you make regarding the pH of a particular part of the digestive system and the type and relative size of the carbon–based molecules that are broken down there?

2. If the optimal pH range of most enzymes is between 6 and 8, is there possibly some advantage to having certain enzymes, such as pepsin, active in the low pH of the stomach?

3. The stomach is a relatively small muscular organ compared to the small intestine that can be as long as six meters. Yet food tends to be held in the stomach and released in only small amounts into the small intestine. What does that suggest to you about where nutrients are absorbed? How do the products of enzyme action described in the table support your thinking?

Name Period Date

CHAPTER 2 | CHEMISTRY OF LIFE

Vocabulary Practice

atom	adhesion	carbohydrate	bond energy
element	solution	lipid	equilibrium
compound	solvent	fatty acid	activation energy
ion	solute	protein	exothermic
ionic bond	acid	amino acid	endothermic
covalent bond	base	nucleic acid	catalyst
molecule	pH	chemical reaction	enzyme
hydrogen bond	monomer	reactant	substrate
cohesion	polymer	product	

A. Who Am I? Choose among these terms to answer the riddles below:

activation energy	enzyme	nucleic acid
bond energy	fatty acid	pH
carbohydrate	hydrogen bond	protein
compound	ion	solution
endothermic	molecule	substrate

1. I am a chemical reaction that absorbs more energy than I release; my products have a higher bond energy than my reactants. ____________

2. I am an attraction between a slightly positive hydrogen atom and a slightly negative atom, such as oxygen or nitrogen. ____________

3. I am the type of molecule that stores genetic information in living things; I include DNA and RNA. ____________

4. I am the amount of energy needed to break a chemical bond; I am also the amount of energy released when a bond forms. ____________

5. I am a certain type of mixture; I am formed when one substance, such as water, dissolves other substances. ____________

6. I allow chemical reactions to take place under the tightly controlled conditions in living things; in almost all cases I am a protein. ____________

7. Chemical bonds hold me together; I always contain atoms of more than one element. ______

8. I am a carbon chain found in many lipids; I can be either saturated or unsaturated. ______

9. I have a positive charge if I lose an electron, or I have a negative charge if I gain an electron. ______

10. I am a reactant an enzyme acts on; I bind to an enzyme's active site. ______

11. Many different types of me have many different functions in living things; I am made of amino acids. ______

12. I measure the hydrogen ion concentration in a solution; when I am low a solution is very acidic. ______

13. I am the amount of energy that is needed for a chemical reaction to start; after I am added, the reaction can take place on its own. ______

14. I am held together by covalent bonds; I can be made of atoms of just one element. ______

15. Carbon, hydrogen, and oxygen are my building blocks; sometimes I am broken down as a source of energy, and sometimes I make up plant cell walls. ______

B. Same or Different? Each pair of words could either describe the same thing or different things. If they can describe the same thing write "same." If they cannot describe the same thing write "different."

______ 1. catalyst/enzyme

______ 2. atom/element

______ 3. cohesion/adhesion

______ 4. monomer/polymer

______ 5. exothermic/endothermic

______ 6. substrate/reactant

______ 7. solvent/solute

______ 8. compound/element

______ 9. acid/base

______ 10. enzyme/protein

VOCABULARY PRACTICE, CONTINUED

C. Compound Word Puzzle Read the phrase and write the word that it most closely describes. Then write another phrase that describes the same word in a different way.

PHRASE 1	WORD	PHRASE 2
smallest basic unit of matter	**Example** atom	contains protons, neutrons, and electrons
enters a chemical reaction	**1.**	
atom that has gained or lost electrons	**2.**	
held together by covalent bonds	**3.**	
dissolves a solute	**4.**	
releases a hydrogen ion into a solution	**5.**	
reactants have a greater bond energy than products	**6.**	
substance that decreases activation energy	**7.**	
includes sugars and starches	**8.**	

Name Period Date

VOCABULARY PRACTICE, CONTINUED

D. Find the Odd Word Put a checkmark next to the word that does not belong and explain why. There may be more than one correct way to answer for some of the word sets.

1. ____ acid Explanation____________________

____ hydrogen bond ____________________

____ base

2. ____ compound Explanation____________________

____ atom ____________________

____ element

3. ____ covalent bond Explanation____________________

____ solution ____________________

____ molecule

4. ____ equilibrium Explanation____________________

____ bond energy ____________________

____ exothermic

E. Situational Vocabulary Circle the letter of the situation that most closely relates to each vocabulary word.

1. **activation energy:** a) kick-starting a motorcycle; b) eating sugary foods
2. **solute:** a) getting lost in a crowd; b) answering a math problem
3. **ionic bond:** a) linking hands in a human chain; b) passing a basketball
4. **lipid:** a) a car's gas tank; b) a car's headlights
5. **amino acid:** a) lemon juice in iced tea; b) one of many beads on a string
6. **covalent bond:** a) linking hands in a human chain; b) passing a basketball
7. **equilibrium:** a) evenly matched tug-of-war; b) sprinters ready to race
8. **atom:** a) multicellular organism; b) unicellular organism

Name Period Date

UNIT 1 | THE ANATOMY OF AN EXPERIMENT

Unit Project

Scientists, regardless of their field of study or the time in which they lived, all have something in common, a "scientific habit of mind." Scientists have a particular approach when trying to answer a question or solve a problem. It involves identifying a question or a problem, developing a proposed answer, often designing and conducting an experiment, analyzing observations and data, and communicating results and conclusions. The knowledge that is gained is shared with the scientific community and the public. Sometimes that knowledge raises new questions and inspires further research.

If a scientist has conducted one or more experiments, it is important that the scientist describe the procedure in a very specific and detailed way so that other scientists can evaluate the work. To appreciate all that goes into designing an experiment, you will be researching a classic experiment and writing a lab report as if you had performed it yourself. You will choose either Francesco Redi's or Louis Pasteur's experiments on spontaneous generation.

WRITING A LABORATORY REPORT

Laboratory reports are often as valuable as the experiment they describe. They communicate information and ensure that an experiment can be replicated. In your lab report be sure to include the following components:

- **Literature Search:** Research what the question or problem was that Redi or Pasteur was studying, what the experiment dealt with, how the experiment was conducted, and what the data and conclusions were. Use at least three documented sources.
- **Problem Statement:** Clearly state what question or problem was under investigation as accurately and completely as possible from the information you have gathered.
- **Experimental Hypothesis:** Write an "if" and "then" statement to show the relationship between the variables. The statement must be specific and testable.
- **Experimental Procedure:** Identify the variables and constants, the experimental and control groups, and the operational definition. Also, provide a detailed materials list and write a detailed, step-by-step procedure for the experiment. Address safety concerns.
- **Analyzing Data:** Organize data in tables or graphs and analyze the results. You may need to supply your own qualitative data, but it should be in keeping with the descriptions of the experiments you find.
- **Conclusion:** State whether or not the hypothesis was supported by the data and observations recorded. Restate the relationship between the variables and explain how the results support your conclusions. Indicate any experimental errors that may have occurred and if they had an effect on the results. Discuss any possible applications and extensions of the research findings.

On the following page you will find the evaluation rubric. Points will be assigned according to how well a task is done, as well as whether you have fully or only partially completed a task. Any task left undone will be assigned a zero. Your teacher will provide you with a timetable for this project.

The Anatomy of an Experiment Evaluation Rubric

Requirements	Maximum Points	Earned Points (teacher to fill in)
Three documented sources are provided.	5	
Question/problem statement is completely and accurately stated.	5	
Experimental hypothesis is written as an "if" and "then" statement, clearly shows relationship between variables, and is testable.	10	
Variables, constants, experimental and control groups, and operational definition are identified.	10	
Complete, detailed list of materials.	5	
Detailed, step-by-step procedure.	10	
Safety concerns addressed.	5	
Observations are organized and recorded.	5	
Data is recorded in tables/graphs. Tables/ Graphs are complete, appropriate, and clearly labeled.	10	
Conclusion includes a thorough description of how data and observations support or reject hypothesis.	10	
Relationship between variables is restated in the conclusion and supported by data.	10	
Experimental errors indicated and effects are discussed.	5	
Applications and extensions of research are discussed.	5	
Report is neat and legible with no spelling or grammar mistakes.	5	
Rubric Score:	100 points	
Extra Credit (given at teacher's discretion):		
Acting as Redi or Pasteur, student presents "their" work to the class.		
	Total Score:	
Teacher's Comments:		

UNIT
1 THE ANATOMY OF AN EXPERIMENT
Unit Project Teacher Notes

Purpose: To learn how to structure and write a formal laboratory report.

Overview: Students will write a formal laboratory report based on either Louis Pasteur's or Francesco Redi's experiments on spontaneous generation. They will:

- Conduct a literature search to find information on the experiment.
- Identify the problem or question each scientist was trying to answer and write the hypothesis, experimental procedure, data and observations, and conclusions for his experiment.
- Write a laboratory report based on the chosen experiment.

Preparation:

- Copy project description, student handout (optional), and rubric.
- Plan timetable.

Project Management:

- Hand out the project description and rubric, reviewing it with the class.
- Assign project at the beginning of Chapter 1. It should take two weeks to complete.
- Chapter 1, Section 1.3, and Student Resource pages R11–R15 contain helpful information relating to scientific thinking, processes, and experimentation.
- Projects are done independently, outside of class.
- You may need to review what constitutes good experimental design and what an operational definition is and how to write a formal hypothesis as an "if" and "then" statement. Make sure that the students know that the hypothesis must be written in a form that is testable and relates the variables.
- You can expect to see qualitative and quantitative data. Most qualitative data will be the appearance of factors in the experiment (for example, if there are maggots on the meat in Redi's experiment). Quantitative data may appear as length of time or the number of maggots (Redi's experiment).
- As the work of Redi was conducted in the 17th century and Pasteur's in the 18th century, descriptions of the experiments will vary and students will not find first-hand accounts. Students will need to use creativity to develop a materials list, procedure, and address safety concerns. You can print out the overview provided and have students look at the equipment in the lab or in a supply catalog to write up the materials list and procedure.
- Possible materials list for Redi's experiment: 6 500-mL Erlenmeyer flasks, beef, 3 1-cm^2 pieces of gauze, 3 rubber bands.
- Possible materials list for Pasteur's experiment: beef or chicken broth, 8 250-mL Erlenmeyer flasks, 2 1" solid rubber stoppers, 4 1" rubber stoppers with one hole, bunsen burner, 2 pieces of bent glass tubing, 2 pieces straight glass tubing.

Differentiation: This project can be adapted for various ability levels within the class.

- **Below Level Students:** Have students research the work of Redi.
- **Pre-AP Students:** In addition to the laboratory report, have students compare the work of both scientists to see how the experiments are related.

Student Handout
Overview of the Experiments

FRANCESCO REDI'S EXPERIMENTS

Redi was attempting to dispute spontaneous generation and the belief that maggots arose spontaneously from decaying meat.

- In his first experiment, Redi placed meat into two different jars. In one jar, the meat was left exposed to the air, while the other jar was sealed with wax. Maggots appeared only in the uncovered jar.
- Redi conducted a second experiment to address the criticism that the lack of fresh air in the wax covered jar would prevent any maggots from growing. Again, one jar was left exposed to the air. The other jar was covered with a layer of gauze so that air could enter the jar. Maggots appeared only in the uncovered jar.

Redi concluded that maggots were only present on meat when a fly was able to land on its surface and lay eggs. Maggots did not appear on the meat in the covered jars because the flies were unable to land on the surface of the meat and lay their eggs. Conclusion: life comes from other life. For reference, a house fly can lay 75–100 eggs at one time. These eggs hatch within a day.

LOUIS PASTEUR'S EXPERIMENTS

Pasteur conducted a series of experiments to dispute the idea that bacteria spontaneously arose from organic matter (nutrient broth).

- Using a regular flask filled with sterile broth that was open to the air, Pasteur found that microorganisms appeared in the flask.
- Using a regular sealed flask with sterilized broth, Pasteur found no microorganisms appeared.
- Pasteur used a swan–neck flask filled with nutrient broth. He boiled the flask to sterilize the broth and left it open to the air. No microorganisms grew.
- Using a regular open flask containing sterile broth, Pasteur used a flame to sterilize the air entering the flask. No microorganisms grew.

Pasteur concluded that microorganisms are present in the air and will grow if they have a supply of nutrients. Microorganisms didn't appear in the sealed flask because the air and any microorganisms it may have contained were unable to reach the nutrient broth. They also didn't appear in the swan–neck flask because the curvature of the flask prevented them from reaching the nutrient broth. Alternatively, by heating the air as it entered the flask, the microorganisms were killed. The only flask that microorganisms appeared in was the one that was open to the air and allowed airborne microorganisms to reach the nutrient broth. Conclusion: life comes from other life.

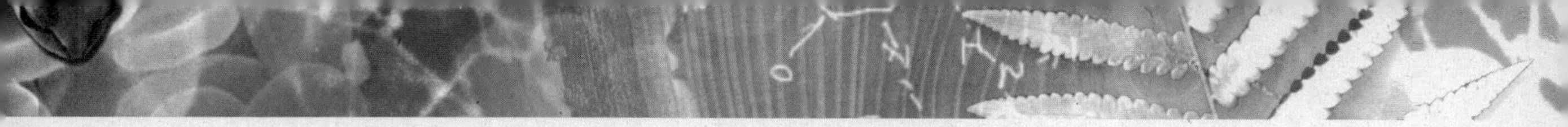

Answer Key

Section 1.1

Study Guide

1. Answers should include different land and aquatic environments, as well as the atmosphere.
2. the variety of life on Earth
3. a type of living thing that can interbreed to reproduce
4. Biodiversity is the variety of life (or species) across the biosphere.
5. It increases from the poles to the equator.
6. made up of cells; uses energy and has metabolism; responds to its environment; can reproduce and grow
7. **Cells:** all organisms are made of one or more cells. **Energy and metabolism:** all organisms need energy for chemical processes, called metabolism. **Response to environment:** all organisms respond to physical factors, called stimuli, in their environment. **Reproduction and development:** all organisms must have the ability to reproduce and pass on their genetic material, or DNA, to their offspring.
8. *Bio-* means "life," and *sphere* means ball-shaped. *Biosphere* means all of the places living things are found on Earth.
9. any individual living thing
10. Metabolism is all of the chemical processes that change substances in an organism.
11. DNA is the genetic material of all organisms, and it is passed on to offspring during reproduction.

Power Notes

Biosphere—Everywhere on Earth where life exists
Biodiversity—Variety of species in a particular area or across the biosphere
Species—A type of living things that can reproduce by interbreeding
Organism—Any individual living thing

1. Cells; All organisms are made up of cells, from single-celled organisms (unicellular) to multicellular organisms.
2. Need for energy—All organisms need chemical energy to carry out their functions including metabolism, which is all of the chemical processes that build up or break down materials.
3. Respond to environment — All organisms must respond to stimuli to survive.
4. Reproduction— Organisms must have the ability to reproduce in order for a species to survive. When organisms reproduce, they pass on their genetic information (DNA) to their offspring.

Reinforcement

1. the biosphere
2. the number of different species in an area or across the entire biosphere
3. based on cells, need energy, respond to stimuli, may reproduce
4. all of the chemical processes that build up or break down materials in an organism

Section 1.2

Study Guide

1. an organized group of related parts that interact to form a whole
2. Sample Answers: body systems; an organ
3. Molecules
4. Different types of cells work together to form a system, such as the circulatory system.
5. Organism
6. Different living things interact with each other and with nonliving parts of their environment.
7. Structure refers to shape and arrangement. Function is how something works or what it does.
8. Sample Answer: Cells in the brain have long branching structures that allow them to send messages long distances.
9. the maintenance of constant internal conditions in an organism
10. Homeostasis keeps cells within a range of conditions

that allow cells to complete their activities.

11. Students' answers will vary but should indicate that negative feedback acts to return a condition to normal.

12. the change in living things over time.

13. changes

14. Adaptations

15. ecosystem

16. homeostasis

Power Notes

Systems—Related parts interact to form a whole. Examples include molecules that function together; cells that work together to perform a function; different organs that work together; different organisms that interact in an ecosystem.

Structure and Function — Function is related to, and dependent upon, structure. Examples include molecular structure and function; cell structure and function; anatomical structure and function.

Homeostasis—Stable internal conditions are maintained through automatic responses and through behavior. Examples include temperature regulation; behavioral responses to external stimuli.

Evolution—Gradual change in genetic makeup of a population of a species. Examples include adaptations in different environments resulting in different species.

Reinforcement

1. living and nonliving things

2. structure is directly related to function at all levels of biology

3. by keeping the internal conditions stable so that cells can function

4. Evolution is the gradual change of the genetic makeup in a population of a species.

Section 1.3

Study Guide

1. use of senses or tools to study the world

2. answers will vary but could include sketches, counting, measurements, etc.

3. observations that are recorded

4. answers will vary but could include examples of qualitative data (colors, sounds, textures) or quantitative data (numeric measurements)

5. proposed answer to a scientific question that is specific and testable

6. answers will vary but could include any of the hypotheses in the chapter or hypotheses that students develop on their own

7. Statistics show whether a hypothesis is supported or rejected by the data.

8. Other scientists can find weaknesses in a scientific report, and peer review helps to maintain honesty in science.

9. observation, forming a hypothesis, testing a hypothesis, analyzing data, evaluating results

10. observational

11. experiments

12. dependent

13. Constants are not manipulated. They are kept the same between experimental conditions or groups.

14. A theory explains a wide range of observations; a hypothesis is a proposed answer to one scientific question.

15. because new scientific evidence can lead to a theory being changed or discarded

16. a proposed answer to a scientific question

17. Students' answers will vary.

Power Notes

Observing—Using the senses or tools to gather information

Forming hypotheses—Ask a question and try to explain observations

Testing hypotheses—Collecting data to support or reject a hypothesis

Analyzing data—Statistical analysis of data to draw conclusions

Evaluating results—Data and conclusions are studied to determine whether they are valid

Experiment
Using independent and dependent variables to find cause-and-effect relationships
Independent variable—The condition that is manipulated in an experiment; the "cause"
Dependent variable—The factor that is measured in an experiment; the "effect"
Constant—Any condition that is kept the same during an experiment; necessary for determining whether the independent variable produces any change in the dependent variable
Theory
A proposed explanation for a wide range of observations and experimental results that is also supported by a wide range of evidence

Reinforcement

1. They propose answers to scientific questions, then test the hypotheses.
2. Independent variables are manipulated by a scientist (the cause); dependent variables are measured by a scientist (the effect).
3. a proposed explanation for a wide range of observations and results that is supported by a large amount of evidence

Section 1.4

Study Guide

1. LMs use light and lenses to magnify images; they can be used with living specimen; they can clearly magnify specimens up to about 1500 times
2. SEMs use electrons and computers to magnify specimens and produce three-dimensional images; they cannot be used to observe living specimens; SEM images are computer colorized; SEMs clearly magnify specimens more than 100,000 times
3. TEMs use electrons and computers to magnify specimens and produce two-dimensional images; they cannot be used to observe living specimens; TEM images are computer colorized; TEMs clearly magnify specimens more than 100,000 times
4. x-rays are absorbed by dense tissues and pass through soft tissues; useful to study skeletal, bone, and tooth structure
5. MRIs use strong magnetic fields to show a cross section of a body part and can be combined by computer to give a complete image of the whole area; can show both soft tissues and dense tissues
6. something that simulates interactions among many different variables and provides scientists with a general idea of how a biological system works
7. to study complex systems and when actual experiments are unsafe, impractical, or unethical
8. a segment of DNA that stores genetic information
9. computer databases store information and allow scientists to search for and retrieve the information
10. A gene is one segment of DNA. A genome is all of an organism's DNA.
11. A genome is all of an organism's DNA. Genomics is the study and comparison of genomes within and among species.
12. Molecular genetics is the study and manipulation of genes on a molecular level.

Power Notes

1. Light and lenses used to magnify specimens; limited magnification, but can be used to study living specimens
2. Deflection of electrons used to magnify specimens; provides high magnification and a three-dimensional black-and-white image that can be colored by computer; cannot be used to study living specimen
3. Electrons passing through a specimen used to magnify specimen; provides high magnification and a two-dimensional black-and-white image that can be colored by computer; cannot be used to study living specimens
4. X-rays passing through tissues used to show dense materials (bones, teeth)

5. Magnetic field used to show all tissues, including soft tissues such as tendons and ligaments
6. Magnetic field that can be used to shown activity in particular brain regions while a subject is performing a task
7. Use of computers to simulate complex systems when actual experiments are not possible
8. Study and manipulation of DNA at the molecular level

Reinforcement

1. Light microscopes use light and lenses, can be used to study living specimens, and can clearly magnify objects up to about 1500 times. Electron microscopes use electrons and computers, cannot be used to study living specimens, and provide much greater magnifications.
2. when they are used to study things that would otherwise be impossible to observe directly
3. A gene is a segment of DNA that stores genetic information; a genome is all of an organism's DNA.

Section 1.5

Study Guide

1. Answers will vary but could include diet, allergies, diseases and disease prevention, genetics, and biochemical effects of drugs.
2. Biology can provide background knowledge, and scientific thinking requires a logical and skeptical review of information.
3. the use and application of living things and biological processes
4. Benefits include improved medical treatments, disease detection, and agricultural production.
5. Risks and ethical concerns include unknown long-term effects, the potential spread of undesirable genes, and privacy concerns.
6. 50
7. unasked
8. We do not have the technology to research and answer some questions.
9. *Trans-* means "across," and *genic* relates to genes; a transgenic organism has genes across, or from, different organisms.
10. *Bio-* means "life," and technology is the application of science for industrial means; biotechnology is the application of life science to the production of materials and techniques.
11. Answers will vary; check student work.

Power Notes

Personal health and lifestyle choices—Issues include diet, obesity, exercise, smoking, alcohol and other drugs, cancer prevention

Environment and society—Issues include interaction between environment and humans, political, legal, economic considerations

Biotechnology

Benefits—food production, medical advances, improved crop yields, detection of genetic disorders, forensic science

Risks—unknown long-term effects of genetically modified crops, safety of GM plants, possibility of decreasing biodiversity

Ethical concerns—privacy of genetic testing, possible discrimination based on genetic screening, ethics of choosing characteristics of children, use of genetic information

Reinforcement

1. Biology can provide an understanding of nutrition, dangerous habits, and diseases.
2. because the environment affects people, and people affect the environment
3. the application and use of living things and biological processes

Chapter 1

Data Analysis Practice

1. The data collected on the amount of time spent feeding in each of the different locations is quantitative data.

2. The observations that describe the difference in size and beak shape between males and females are qualitative. Also, the difference in the shape of flower from which the males and females obtain nectar is qualitative.
3. Size and shape differences in the birds and flowers could be changed into quantitative data if the sizes and shapes were measured.

Pre-AP* Activity

EVOLUTION AND NATURAL SELECTION

1. Over the course of generations, mutations could have accumulated that made some of the insects increasingly thornlike in shape. Consequently, birds preying on the insects would have an increasingly difficult time finding them among the thorns. Because the birds were preying mostly on the "non-thorny" insects, the thornlike ones were more likely to survive and reproduce. The result was a population with a high percentage of thornlike bugs. This trend could continue, with the trait for increasingly thornlike body shape being passed on. Eventually the insect's shape became a near match to the actual thorns and the trait was characteristic of every insect in the population.
2. DNA
3. Those organisms that are not well suited to survive and reproduce in a particular environment do not pass on their traits to offspring.
4. Students' drawings should show a gradual extension of the insects' body behind the head.

Pre-AP Activity

OPTICS OF A MICROSCOPE

1. inverted
2. Students' ray diagrams should show that the image of the object at point B is over twice the size of the image of the object at point A. The difference is that the closer the object is to a lens, the wider the angle of the light rays coming into the lens, hence the bigger the image.
3. upright
4. Students' diagrams should show an enlarged real image just to the right of F'_e. This serves as the object of the eyepiece, which produces a greatly enlarged virtual image to the right of F_o. The object needs to be placed behind the focal point of the objective lens because a real image is needed to serve as the object of the second lens. Both images are inverted, but the second one is a virtual image.

Vocabulary Practice

A. Categorize Words

1. organism, L; cell, L; species, L; transgenic, B; biotechnology, T; molecular genetics, T
2. homeostasis, I; biosphere E; system, B; metabolism, I; ecosystem, E; biology, B
3. biodiversity, G; organism I; cell, B; adaptation, B; biosphere, G; evolution, G

B. Vector Vocabulary

1. using tools and senses to study the world
2. many support, and are accounted for by
3. can lead to
4. a proposed explanation for a wide range of observations supported by a large amount of evidence
5. a proposed answer for a scientific question
6. tested through
7. study of independent and dependent variables to find cause-and-effect relationships
8. scientists manipulate
9. scientists measure
10. scientists control
11. condition that is changed to observe an effect on the dependent variable
12. factors that are observed; affected by independent variable
13. conditions that are kept from changing during an experiment
14. collected as
15. observations that are recorded and analyzed to test a hypothesis

C. Stepped-Out Vocabulary

1. scientific study of all organisms; helps to make informed decisions; still has many unanswered questions
2. tool that provides an enlarged image of an object; led to discovery of cells; electron microscopes provide greater magnifications than light microscopes
3. change in living things over time; change in genetic makeup of a population; accounts for unity and diversity of life
4. an inherited trait that gives an advantage to an individual organism; passed on to future generations; different adaptations in different environments
5. the genetic material in all organisms; instructions for growth and development; passed on through reproduction
6. segment of DNA that stores genetic information; studied and changed through molecular genetics; faulty ones can be replaced
7. study and comparison of genomes; computer databases store the information; entire DNA sequences from organisms

D. Words in Context

1. too small to be seen
2. maintenance of constant conditions
3. the cause
4. segment of DNA
5. inherited trait
6. kept the same
7. explains many observations
8. both living and nonliving things
9. near the equator
10. a type of living thing that can reproduce
11. it has genes from different species
12. propose an answer to a question

Section 2.1

Study Guide

1. An element is a certain type of atom.
2. Sketch should resemble one of the illustrations in Figure 2.2 on page 37. Nucleus should be in the center with protons and neutrons. Electrons should be labeled in a ring surrounding the nucleus.
3. A compound is composed of atoms of different elements.
4. an atom that has gained or lost one or more electrons
5. the number of protons no longer equals the number of electrons
6. Sketch should resemble the first illustration in Figure 2.3, with one atom losing an electron (becoming a positive ion) and the other accepting that electron (becoming a negative ion).
7. A bond that forms between two atoms that share a pair of electrons.
8. the number of electron pairs that it needs to share to fill its outer energy level
9. molecule
10. compound
11. element
12. ion
13. Ionic bonds form between oppositely-charged ions; covalent bonds form when atoms share a pair of electrons.

Power Notes

Atom

smallest basic unit of matter

Energy levels—regions around a nucleus in which electrons are located

Outermost energy level—electrons in the outermost energy level determine how atoms form ions or form covalent bonds

Element 1—hydrogen

Element 2—oxygen

1. electron
2. nucleus
3. protons
4. neutrons

Compound

made of atoms of two or more elements bonded together

Ions—atoms that gain or lose one or more electrons

Ionic bonds—form between oppositely charged ions due to electrical attraction

Positive ions—formed when atoms lose electrons, tend to be formed by atoms with few outer energy level electrons

Negative ions—formed when atoms gain electrons, tend to be formed by atoms with nearly full outer energy level
Covalent bonds—formed when atoms share one or more pairs of electrons
Molecules—atoms bonded together by covalent bonds; not necessarily a compound

Reinforcement

1. nucleus (protons and neutrons) and electrons
2. the number of protons in the atom's nucleus
3. An ionic bond forms through the electrical force between oppositely charged ions; a covalent bond forms between atoms that share pairs of electrons.

Section 2.2

Study Guide

1. a molecule with a slightly positively charged region and a slightly negatively charged region
2. The oxygen nucleus pulls the electrons in the molecule more strongly than the hydrogen atoms. The hydrogen atoms have a slight positive charge and the oxygen atom has a slight negative charge.
3. an attraction between a slightly positive hydrogen and a slightly negative atom
4. between the slightly positive hydrogen atom of one water molecule and the slightly negative oxygen atom of another water molecule
5. A large amount of energy is needed to produce an increase in water temperature; water resists changes in temperature.
6. Water molecules "stick" to each other.
7. Water molecules "stick" to other substances.
8. A solvent is present in greater concentration and dissolves the solute.
9. ions and polar molecules
10. nonpolar molecules, such as fats and oils
11. **Effect on H^+ concentration:** acids increase H^+ concentration; bases decrease H^+ concentration
Effect on pH: acids lower pH.; bases raise pH.
12. Sketch should resemble Visual Vocab on page 42 of the text. Solution: homogeneous mixture of substances; solvent: substance present in greater amount in which other substances dissolve; solute: substance that dissolves in a solvent.

Power Notes

Polar molecules—molecules that have regions with slight electrical charges due to uneven pull on electrons
Nonpolar molecules—molecules without charged regions due to equal pull on electrons

1. hydrogen bonds
2. formed by attraction between slightly positive hydrogen atom and slightly negative (oxygen) atom

Atom 1—oxygen; slightly negative
Atom 2—hydrogen; slightly positive

1. High specific heat—large amount of energy needed to produce an increase in temperature
2. Cohesion—water molecules "stick" to each other
3. Adhesion—water molecules "stick" to other substances

Solutions—a mixture that is the same throughout (homogeneous)
Solvents—substance present in greatest concentration; dissolves other substances
Solutes—substance present in lower concentration; dissolves in solvent
Acids—release H^+ ions in solution; high H^+ concentration; low pH (<7)
Neutral—neither acidic nor basic; pH of 7
Bases—remove H^+ ions from solution; low H^+ concentration; high pH (>7)

Reinforcement

1. Slightly positively charged hydrogen atoms are attracted to slightly negatively charged atoms.
2. solvent and solute

3. the hydrogen ion concentration in a solution

Section 2.3

Study Guide

1. Carbon atoms are the basis of the molecules that make up most living things.
2. Carbon atoms can form covalent bonds with up to four other atoms, including other carbon atoms.
3. Sketches should resemble Figure 2.10 on page 44.
4. broken down as a source of chemical energy; part of cell structure
5. sugars, starches, cellulose
6. broken down as a source of chemical energy; part of cell structure
7. fats, oils, phospholipids
8. many functions, including movement, transport, chemical catalysts
9. enzymes, hemoglobin
10. store genetic information, build proteins
11. DNA, RNA
12. the order of amino acids and interactions between amino acids (hydrogen bonds and sulfur-sulfur bonds)
13. nucleotides, which are composed of a sugar, a phosphate group, and a nitrogen-containing base
14. A monomer is a single subunit; a polymer is a molecule made of many monomers.

Power Notes

Monomer—small molecule that is a single unit in a much larger molecule
Polymer—large molecule (macromolecule) formed from many monomers bonded together

1. Functions—broken down to provide a source of energy; make up plant cell walls
   Monomer—glucose
2. *Lipid*
   Functions—broken down to provide a source of energy; make up cell membranes; used to make steroid hormones
   Molecule—phospholipid; phosphate group head; nonpolar fatty acid tails
3. *Protein*
   Functions—many, including enzymes, oxygen transport, and muscle movement
   Polymer—polypeptide (or protein) Monomer—amino acid

   Nucleic acid
   Types—DNA and RNA
   Functions—store genetic information; build proteins

Reinforcement

1. broken down as a source of chemical energy; part of cell structure
2. sugars, starches, cellulose
3. broken down as a source of chemical energy; makes up cell membranes
4. fats, oils, phospholipids
5. many functions, including movement, transport, chemical catalysts
6. enzymes, hemoglobin
7. store genetic information, build proteins
8. DNA, RNA

Section 2.4

Study Guide

1. Reactants are on the left side of the equation. They are substances that are changed during a chemical reaction. Products are on the right side of the equation. They are substances made by a chemical reaction.
2. the addition of energy to the reactants
3. the energy required to break a bond between two atoms
4. one arrow on top of another, pointing in opposite directions (also accept a sketched version of answer)
5. when a reaction takes place at equal rates in both directions
6. bond energy
7. energy, activation energy
8. lower, released
9. higher, absorbed
10. Sample answer: A chemical reaction changes reactants into products.
11. Sample answer: the energy it takes to get out of bed in the morning before you can start your day
12. The reactants and products are formed equally; the reaction is in a balanced state.
13. Energy goes out of an exothermic reaction and

goes into an endothermic reaction.

Power Notes

1. reactants; substances changed by a chemical reaction
2. direction of the reaction
3. products; substances formed during a chemical reaction
4. reactants
5. activation energy
6. total energy released
7. products
8. products
9. activation energy
10. total energy released
11. products

Bond energy
the amount of energy needed to break a bond, or the amount of energy released when a bond forms
Chemical equilibrium
reversible reactions can occur in both directions (reactants to products, or products to reactants, depending on reactant and product concentrations); when reactants and products are formed at the same rate, a reaction is at equilibrium
Exothermic
more energy released than absorbed; energy usually released as heat and/or light; reactants have higher bond energy than products
Endothermic
more energy absorbed than released; products have higher bond energy than reactants

Reinforcement

1. reactants and products
2. the amount of energy necessary to start a chemical reaction
3. Exothermic reactions release more energy than they absorb, but endothermic reactions absorb more than they release.

Section 2.5

Study Guide

1. the amount of energy required to start a chemical reaction
2. Catalysts decrease activation energy for a chemical reaction; catalysts increase the rate of a chemical reaction; catalysts are neither reactants nor products because they are not changed or used up.
3. less
4. **Why enzymes are necessary:** because reactions in organisms have to occur at a low temperature (body temperature), with low concentrations of reactants, and at a high rate; **Important factors in enzyme structure:** order of amino acids and hydrogen bonding between the amino acids; **How structure affects function:** if structure changes, substrates will not be able to bind to an enzyme's active sites; **Lock-and-key model:** only certain substrates bind to an enzyme's active sites; when bound to the enzyme the chemical reaction can occur.
5. The enzyme's shape changes slightly, which strains the bonds inside the substrate. The strain on the bonds weakens them.
6. A catalyst "dissolves" or "gets rid of" some of the activation energy needed to start a reaction.
7. Specific substrates fit exactly into the active sites for specific enzymes, in a similar way that only a certain key will open any given lock.

Power Notes

1. decreases activation energy and increases reaction rate
2. is not used up during a reaction and does not alter the equilibrium of the reaction

Enzymes—catalysts for chemical reactions in living things; almost all are proteins
Homeostasis and enzymes — large changes in temperature or pH can cause enzymes to break down (hydrogen bonds break); stable conditions necessary for an organism's enzymes to function properly
Enzyme structure and function — altered enzyme structure prevents the enzyme from functioning properly

Substrates—reactants acted upon by an enzyme; bind to an enzyme's active sites
Lock-and-key model—only particular substrates will bind to particular enzymes; when bound, enzymes weaken bonds within substrates and allow reactions to occur

Reinforcement

1. a substance that decreases the activation energy, and increases the rate of a chemical reaction
2. in living things
3. Substrates bind to an enzyme, which catalyzes a chemical reaction. The enzyme brings substrates together and weakens their bonds, which allows the reaction to take place faster and with less activation energy.

Chapter 2

Data Analysis Practice

1. The independent variable is the amount of antacid (mg). The dependent variable is the amount of time it takes for the antacid to neutralize 1 mL of hydrochloric acid.
2. The greater the amount of antacid, the less the amount of time that is needed to neutralize the acid.

Pre-AP Activity

SURFACE TENSION AND COHESION

Experiment results: Students should be able to put about 40 drops of water, 24 drops of rubbing alcohol, and 14 drops of detergent solution on the penny before they spill over.

1. Diagrams should show that molecules in the interior of the pile have attractive forces in all directions, whereas molecules at the surface have a net force toward the interior.
2. Water should rank highest, rubbing alcohol second, and detergent solution last. Numbers of drops will vary by the size of the dropper and other factors.
3. Rubbing alcohol forms fewer hydrogen bonds between its molecules than water does.
4. Students may suggest a difference in drop size produced by different droppers, a difference in how clean the pennies were, or differences in how students handle the droppers.

Pre-AP Activity

ENZYME ACTION

1. *Sample Answer:* pH value for the mouth is neutral and value for the small intestine is only slightly acidic. These environments are more suited to the digestion of carbohydrates such as sugars and starches. The pH value for the stomach is acidic and is better at digesting proteins. Larger molecules tend to be broken down in the stomach and then passed along to the intestine.
2. *Sample Answer:* Having enzymes, like pepsin, active only in an acidic environment effectively limits its activity to an area of the body where large amounts of materials are meant to be broken down so nutrients can be absorbed.
3. *Sample Answer:* The nutrients are being absorbed in the small intestine after the stomach has broken them down somewhat. The types of end products of enzyme action in the small intestine are the smaller monomers that are the building blocks used by cells.

Vocabulary Practice

A. Who Am I?

1. endothermic
2. hydrogen bond
3. nucleic acid
4. bond energy
5. solution
6. enzyme
7. compound
8. fatty acid
9. ion
10. substrate
11. protein
12. pH
13. activation energy
14. molecule
15. carbohydrate

B. Same or Different?

1. same
2. same
3. different
4. different

5. different
6. same
7. different
8. different
9. different
10. same

C. Compound Word Puzzle

1. reactant; changed by a chemical reaction
2. ion; positively or negatively charged
3. molecule; may be an element or a compound
4. solvent; present in greatest concentration in a solution
5. acid; lowers pH of a solution
6. exothermic; reaction that releases energy
7. catalyst; substance that increases reaction rate
8. carbohydrate; can be broken down for energy

D. Find the Odd Word

1. hydrogen bond; the other two terms refer to hydrogen ion concentration in a solution
2. compound; atoms and elements are made of just one substance, compounds contain more than one substance
3. solution; molecules are held together by covalent bonds
4. equilibrium; bond energy determines whether a reaction is exothermic

E. Situational Vocabulary

1. a
2. a
3. b
4. a
5. b
6. a
7. a
8. b